HYDROPOWER PLANTS

HYDROPOWER PLANTS

A Contemporary Guide for Hydropower Plants in the Global Scope

ŠTEFAN TKÁČ
Slovak-Taiwan Hydropower Research Laboratory, Košice, Slovak Republic

LUKÁŠ SEČKA
Department of Architecture, Faculty of Arts, Technical University of Košice, Košice, Slovak Republic

ERIC DENG
Slovak-Taiwan Hydropower Research Laboratory, Košice, Slovak Republic

ELSEVIER

Elsevier
Radarweg 29, PO Box 211, 1000 AE Amsterdam, Netherlands
125 London Wall, London EC2Y 5AS, United Kingdom
50 Hampshire Street, 5th Floor, Cambridge, MA 02139, United States

ISBN: 978-0-12-824520-0

For Information on all Elsevier publications
visit our website at https://www.elsevier.com/books-and-journals

Publisher: Priscilla Braglia
Acquisitions Editor: Ruchi Bhardwaj
Editorial Project Manager: Harshita Manchanda
Production Project Manager: Anitha Sivaraj
Cover Designer: Gopalakrishnan Venkatraman

Typeset by MPS Limited, Chennai, India

Contents

1

Introduction

The value of water

Before proceeding any further, it is important to understand the technical significance of the element we are about to discuss. In fact, water was seen as a symbol of life long before it became a driving force behind human hydraulic inventions. Its power places it on an imaginary pedestal across all nations, cultures, and religions, stretching from the Theertha in Hindu rituals, Transcendence in Christianity, to a simple symbol of purity in Shintoism.

Water (H2O) is a common molecule not limited to Earth. It has a presence beyond our planet, found throughout the universe. Ice forms from water on various planets and moons, such as Europa and Enceladus, as well as in comets. There are also beliefs that hidden oceans lie beneath the icy surfaces of moons in different galaxies. Water vapor in the interstellar medium creates massive clouds containing more water than Earth's oceans—trillions of times more (AAAS, 1998; Space.com Staff, 2011). Water is vital for astrobiology because it has unique properties, such as being a universal solvent and having excellent thermal stability, which are crucial for the development and support of life as we know it. When searching for other forms of life on exoplanets, scientists specifically look for traces of water. It is believed by scientists that the comets and asteroids that impacted Earth may have brought water along with them. On the other hand, a portion of Earth's water that originates from the interior of our planet may have been transported to Earth via the solar wind (Sci.News, 2021). The adaptability of water is enhanced by its ability to exist as a gas (water vapor), in liquid form (water), and as a solid (ice). An important aspect of the adaptability of water is related to enthalpy (hidden heat, also known as latent heat, which is added or removed from materials during phase changes, such as condensation, vaporization, freezing, and melting). Due to the very strong hydrogen bonds between molecules of water, the amount of enthalpy associated with these changes is much higher than for many other materials. Therefore, the concept of enthalpy is very important when looking at the role of water in our environment from two perspectives: macro (e.g., regulating the

Hydropower Plants. DOI: https://doi.org/10.1016/B978-0-12-824520-0.00006-X

climate, maintaining ocean temperatures) and micro (e.g., stabilizing the body temperature of mammals).

> *"Since water is the carrier of life, it also carries energy for life. On the contrary, if uncontrolled or disturbed, it is capable of rapid, massive destruction or slow, steady erosion."*
>
> **(Štefan Tkáč, 2025, while preparing this publication)**

Water, human, power nexus

Water is natural to humans and their development activities from the beginning. It bonds nature and the human body. The human fetus perceives water as a natural environment, significantly contributing to the diving or innate swimming reflexes of newborns during the first 6 months. 60% of the adult human body is water. If there is a deficiency, homeostasis is immediately disrupted, and the body becomes vulnerable to deterioration, degradation, and eventually death. The water scarcity described in the United Nations' Sustainable Development Goal no. 6 is highly relevant to geography, nation, weather, politics, and solidarity. However, it holds much more power than one might imagine.

Among the classical elements described in ancient cultures, human activity has always been most closely linked to water. In fact, our first cities were built next to water. Although they were suppressed during the postindustrial era of large urban developments, recent landscape theories and practices are exploring solutions to bring water back to the city and keep it there (water-sensitive urban design, Sponge City Theory, storm water management, small vs. large water cycles in the new water paradigm (Kravčík et al., 2007),nature-based design, etc.).

Currently, water covers up to 75% of our planet's surface, hidden in various forms. Oceans make up 97% of Earth's surface. Only 3% is freshwater, with 98.8% stored in ice and ground (70% in icebergs, roughly 29% in groundwater). Excluding ocean-related hydropower systems, the remaining water forms, such as lakes, irrigation canals, fishponds, reservoirs, and rivers, can be regarded, for this book, as different types of theoretical hydropower potential (Fig. 1.1).

Where there is water, there is life; where there is water, there could be development. Our cities were originally built next to water bodies for multiple reasons (transportation and trade, agricultural resources, water supply and sanitation, defense), one of which is also energy (McDonald et al., 2014; Winiwarter et al., 2016). The current trend in landscape and urban design is to reintroduce water into urban areas, especially where it was misplaced, forgotten, or

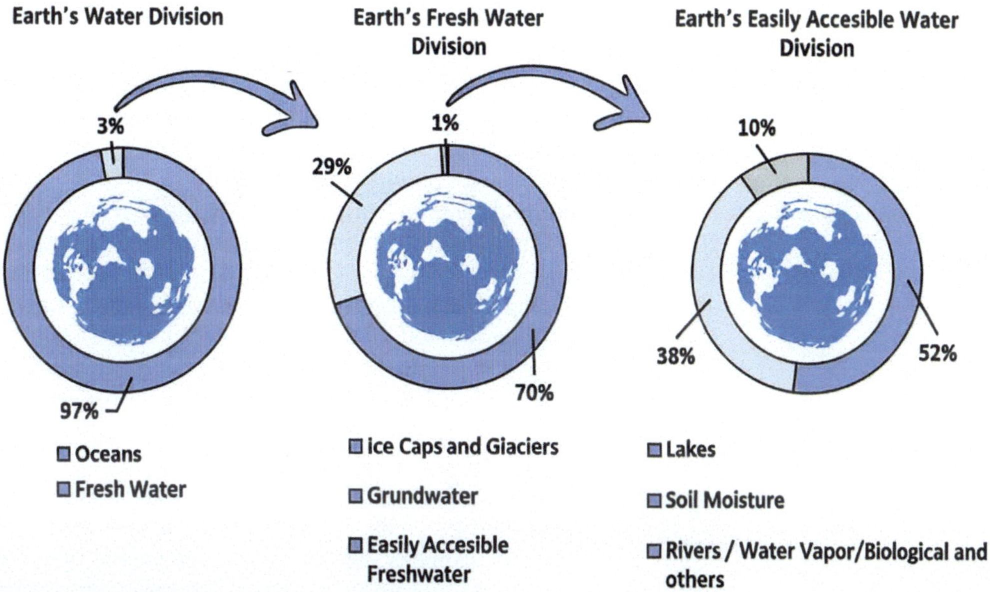

Figure 1.1 Simplified diagram of Earth´s water di/ vision.
Picture Based on Australian Environmental Education. (2021). Water is found everywhere on Earth, so why is it important? *Retrieved December 16, 2025, from https://www.australianenvironmentaleducation.com.au/education-resources/what-is-water/.*

unmaintained, or displaced by the expansion of space for personal transportation. In Architecture, the term Blue Architecture (Muller, 2021) refers to a Water-Centric Design Philosophy that incorporates strategies such as water-sensitive design (WSD) (Sedrez et al., 2021)

The global demand for renewable energy sources (RES) is once again driving advances in research on water bodies. The previously declining development of hydropower systems is gaining renewed attention, now seen as a mature, well-established technology that has evolved over time in the form of a leading RES technology. It also includes versatile, decentralized, and acupuncture-like energy solutions, and an adaptable hybrid platform for relatively new and fluctuating gains from the RES industry, forming the basis of the new energy mix.

Global hydropower facts and trends

Major buy-downs (ENERGO-PRO in the Czech Republic, Google and Brookfield Renewable in the US and Canada, JSW Neo Energy in India, Statkraft in Norway, etc.) dominate the renewable generation sector—-especially hydropower—surpassing other technologies in terms of total installed and operational capacity of RES-powered plants, as well as the amount of electricity produced annually.

Hydropower is believed to have enormous potential worldwide, based on the conventional principle of using water flow and elevation. This excludes ocean power-based systems, which have an estimated theoretical potential exceeding 180 PWh per year, with exploitable potential ranging from 40 to 60 PWh per year (Drishtiias, 2019). The total hydropower potential is estimated at around 40,000 TWh/yr, or roughly 52 PWh/year (52,000 TWh/year), accounting for about a third of global energy consumption. However, based on feasibility studies conducted so far, only an estimated 14,000 TWh/year of hydropower is practically available, and currently, only 7000 TWh/year would be economically viable under present market conditions. The majority of hydropower potential—approximately 90%—remains accessible in developing regions where resources have not yet been fully harnessed (Hoes et al., 2017; Yüksel, 2009).

The total installed capacity of hydropower is greater than the combined installed capacity of all other renewable generation technologies. In 2024, hydropower plants produced an estimated 4500 TWh (terawatt-hours) per year, equivalent to approximately 14% of global energy generation.

The rapid growth of hydropower has made it the single largest renewable electricity source (Darmawi et al., 2012) and the 3rd-largest overall electricity generation source worldwide (IEA, 2025a). Hydropower has become a global energy resource, benefiting people in both developed and developing nations. Hydropower can provide a large variety of energy requirements in rural and urban areas throughout the world. In some cases, there is a single large hydropower facility interconnected to a large grid; elsewhere, smaller decentralized design—developed with lower costs but higher yield—will operate at higher overall efficiency when integrated with multiple other variable renewable energy sources. All these developments cumulatively demonstrate that the continued growth and modernization of hydropower installations, combined with new installations, further prove the importance of hydropower within the transition to a sustainable global energy system.

According to IEA (IEA, 2025b), hydropower currently produces more electricity than all other renewable technologies combined and is expected to remain the world's largest source of renewable electricity in the 2030s. In 2023, the global installed capacity of all types of hydropower plants was approximately 1416 GW, while through 2024 it was approximately 1443 GW (IHA, 2024). By the end of 2024, the total global installed hydropower capacity reached 1450 GW. Hydropower remains the largest source of renewable electricity (Kaldellis, 2025) worldwide and the third-largest source of overall power generation, boasting over 1416 GW of installed capacity and producing up to 4500 TWh annually in 2024–2025, which represents approximately 14–15% of global electricity generation. From 1450 GW, pumped storage hydropower accounts for as much as 200 GW (189 GW between 2023 and 2025), equivalent to roughly 94–95% of all active global energy storage capacity.

Building law regulations, environmental considerations, and increased demand for renewable energy are shaping the direction of hydropower development, with a significant focus on developing sustainable solutions to produce green energy. As a result, research into smaller hydropower plants and alternative energy sources has been pursued in more detail. Large-scale hydropower plants continue to evolve worldwide, but especially in developing countries, such as China, ASEAN member nations, Africa, and the Middle East. Therefore, the expectation of a modified and/or accelerated forecasted development pattern worldwide for installed hydropower capacity does not appear unreasonable (Table 1.1).

On the contrary, with advances in precision and digitalization, more sophisticated autonomous and resilient hydropower technology enables developed countries to focus on refurbishing and retrofitting existing hydropower plants to meet energy demands and boost efficiency. Therefore, the push for a significant increase in globally installed capacity appears to follow a decreasing rather than increasing trend, which is necessary to achieve the 2030 (1650 GW) and 2050 (2800 GW) net-zero targets (IHA, 2024).

Hydropower as a foundation for the renewable energy era

It is generally recognized that RES are versatile, can be decentralized, and utilize energy sources found on-site rather than transport them from afar. This explains their recent widespread use, especially in areas where large-scale systems would be economically unfeasible or ineffective. Using the analogy that "water is formless, shapeless—adaptable, etc." it is not surprising that hydropower technology is closest to humans and nature, making it an adaptable form of energy conversion. When enhanced with digital monitoring tools, operational modes, and predictive models for water-to-grid scenarios, it can operate and adapt to changing environmental conditions, either as a single system or in hybrid models, store energy for short or long periods, stabilize the grid, or serve as the foundation for a developing RES-influenced energy mix in each country (Quaranta et al., 2025)The transition to the RES era without hydropower is not possible (Gatta et al., 2025).

Table 1.1 Scenarios for a global installed hydropower capacity growth for a 2000 GW target (data based on average annual growth in 2024 by IHA).

	Linear growth **based on historical data**	Moderate growth **slightly exceeding historic data**	High growth **in optimistic scenario**
At the end of 2024, the overall global installed capacity of hydropower was 1450 GW, when would it reach **2000 GW**	~25 GW/ year (based on 2024 data by IHA) 2046 (22 years)	Average 35[a] GW/year 2040 (16 years)	Average 45[b] GW/year 2036 (12 years)

Note: Used formula (2000 GW − 1450 GW=550 GW) 550 ÷ 25=22 years.

[a] Given the assumption that the additions accelerate modestly due to projects in development (~1075 GW pipeline) being completed, and the overall legislative environment, along with financing, improves in developing regions.

[b] Assumption based on strong policy support, faster financing, and grid-integration drivers (especially pumped storage development).

The book aims

As a result, this book is intended to serve as a helpful resource for people of all types—including engineers, researchers, planners, architects, policy makers, and students—who will find many valuable sources of information about hydropower in today's renewable energy industry. Engineers and researchers will learn about turbine types, types of hydropower plants, new ways technology can enhance hydropower efficiency, and how to address issues related to retrofitting existing facilities, environmental impact, system integration, etc. Planners and architects will have helpful guidance on organizing the development of space for hydropower facilities, incorporating hydropower into its broader context or surroundings, and using architecture to express hydropower.

The importance of understanding how hydropower facilities function, applying to their particular geographic area; using the specific geography of a site as part of their design; applying to the location of the facility and how it relates to surrounding communities; and having an understanding of the social and cultural history of that geographic area as part of the development process for hydropower. The first section of this book provides an overview of international hydropower development and offers decision-makers and stakeholders information on current policies governing hydropower development, including key sustainability issues. Finally, for students and newcomers to hydropower, this book provides a well-structured educational resource that connects the historical foundations of hydropower with its current practices and offers future perspectives on its evolution in a changing world, positioning hydropower as not only an established technology but also a developing and adaptable component of a low-carbon, resilient energy system.

List of abbreviations

ASEAN	The Association of Southeast Asian Nations (1967) *(Brunei, Cambodia, Indonesia, Laos (Lao PDR), Malaysia, Myanmar, Philippines, Singapore, Thailand, Timor-Leste, and Vietnam)*
IEA	International Energy Agency (1974)
IHA	International Hydropower Association (1995)
PSH	Pumped Storage Hydro
PW / PWh	Peta watt / Peta watt-hours(unit of energy)
RES	Renewable Energy Resources/Sources
TW / TWh	Terawatt / Terawatt-hours
WSD	Water Sensitive Design
yr	year

References

AAAS. (1998). A cloud of water in interstellar space, Cornell University—EurekAlert!—American Association for the Advancement of Science. https://www.eurekalert.org/news-releases/1015461?utm.

Australian Environmental Education. (2021). Water is found everywhere on Earth, so why is it important? Retrieved December 16, 2025. https://www.australianenvironmentaleducation.com.au/education-resources/what-is-water/.

Darmawi, N., Sipahutar, R., Bernas, S. M., & Imanuddin, M. S. (2012). Renewable energy and hydropower utilization tendency worldwide. *Renewable and Sustainable Energy Reviews, 17*, 213–215. https://doi.org/10.1016/j.rser.2012.09.010.

Drishtiias. (2019). Ocean Energy. Retrieved December 16, 2025. https://www.drishtiias.com/to-the-points/paper3/ocean-energy#:~:text=such%20as%20fishing.-,Potential,to%20be%20about%2040%2C000%20MW.

Gatta, F. M., Geri, A., Lauria, S., Maccioni, M., & Nati, L. (2025). The path towards decarbonization: The role of hydropower in the generation mix. *Energies, 18*(19), 5248. https://doi.org/10.3390/en18195248.

Hoes, O.a C., Meijer, L. J. J., Van Der Ent, R. J., & Van De Giesen, N. C. (2017). Systematic high-resolution assessment of global hydropower potential. *PLoS One, 12*(2), e0171844. https://doi.org/10.1371/journal.pone.0171844.

IEA. (2025b). Hydroelectricity. Retrieved December 15, 2025. https://www.iea.org/energy-system/renewables/hydroelectricity.

IEA. (2025a). Retrieved December 15, 2025. https://www.iea.org/energy-system/renewables/hydroelectricity.

IHA. (2024). 2024 World Hydropower Outlook Opportunities to advance net zero. Retrieved December 15, 2025. https://acrobat.adobe.com/id/urn:aaid:sc:us:6ba5f8fc-5ad3-4d52-a83c-0931ce5fa119?viewer%21megaVerb=group-discover.

Kaldellis, J. K. (2025). Long-Term analysis of Hydropower's pivotal role in sustainable future of Greece. *Energies, 18*(9), 2214. https://doi.org/10.3390/en18092214.

Kravčík, M., Pokorný, J., Kohutiar, J., Kováč, M., & Tóth, E. (2007). *Water for the Recovery of the Climate - A New Water Paradigm*. Žilina: Municipalia a.s. and TORY Consulting a.s, 4–94.

McDonald, R. I., Weber, K., Padowski, J., Flörke, M., Schneider, C., Green, P. A., Gleeson, T., Eckman, S., Lehner, B., Balk, D., Boucher, T., Grill, G., & Montgomery, M. (2014). Water on an urban planet: Urbanization and the reach of urban water infrastructure. *Global Environmental Change, 27*, 96–105. https://doi.org/10.1016/j.gloenvcha.2014.04.022.

Muller, B. (2021). *Blue Architecture. University of Texas Press eBooks*. https://doi.org/10.7560/325100.

Quaranta, E., Aliki, G., Letout, S., Mountraki, A., Ince, E., Gea Bermudez, J., & Schmitz, A. (2025). *Hydropower and Pumped Storage Hydropower in the European Union Status Report on Technology Development, Trends, Value Chains and Markets*. Luxembourg: Publications Office of the European Union, 2025. https://doi.org/10.2760/3389272.

Sci.News. (2021). Part of Earth's water came from the sun, new study suggests. *Sci.News: Breaking Science News*. https://www.sci.news/astronomy/solar-water-10316.html?utm.

Sedrez, M., Xie, J., & Cheshmehzangi, A. (2021). Integrating water sensitive design in the Architectural Design Studio in China: challenges and outcomes. *Sustainability, 13*(9), 4853. https://doi.org/10.3390/su13094853.

Space.com Staff. (2011). Astronomers find largest, oldest mass of water in universe. In *Space*. https://www.space.com/12400-universe-biggest-oldest-cloud-water.html?utm.

Winiwarter, V., Haidvogl, G., Hohensinner, S., Hauer, F., & Bürkner, M. (2016). *The long-term evolution of urban waters and their nineteenth century transformation in European cities*. A comparative environmental history. *Water History, 8*(3), 209–233. https://doi.org/10.1007/s12685-016-0172-z.

Yüksel, I. (2009). Dams and hydropower for sustainable development. *Energy Sources Part B Economics Planning and Policy, 4*(1), 100–110. https://doi.org/10.1080/15567240701425808.

2

History

Introduction

The history of using the power of water

The history of hydropower is, at its core, the history of humanity's ability to transform the stored potential of the water into usable mechanical and, later, electrical energy. Since the earliest settlements, water has been perceived not only as a life-sustaining resource but also as a source of power embedded in the landscape of villages, towns, and, increasingly, urban environments. Water bodies may be understood as either static or dynamic, and as naturally formed or artificially created, each category offering distinct energetic characteristics and technological opportunities.

Naturally occurring static water bodies include lakes, ponds, tarns, and wetlands, while their artificial counterparts comprise fishponds, reservoirs, and water basins formed through industrial, mining, or cooling processes. Dynamic water bodies, by contrast, are represented by brooks, streams, rivers, waterfalls, and geysers, whose movement provides a higher kinetic energy potential. Human-made dynamic systems, such as canals and diversion channels, have historically served agricultural, industrial, and transportation functions while simultaneously enabling energy extraction. Together, these forms constitute hydraulic power, defined as a combination of potential and kinetic energy with a quantifiable output that can be harvested, transmitted, and transformed through hydropower technologies.

Throughout history, the higher kinetic potential of flowing water drove the development of increasingly sophisticated energy-harvesting devices. Early water wheels, adapted to local hydraulic conditions, marked the first systematic attempts to convert water energy into mechanical work. Over time, these open and often exposed technologies evolved into enclosed and highly efficient water turbines, forming the technological backbone of modern hydropower generation.

Beyond conventional turbine-based systems, the broader history of water energy utilization also encompasses alternative and emerging technologies. These include the harvesting of thermal gradients

Hydropower Plants. DOI: https://doi.org/10.1016/B978-0-12-824520-0.00001-0

through Ocean Thermal Energy Conversion (OTEC), the capture of kinetic energy from waves and tides, and the exploitation of salinity gradients in osmotic power plants. In parallel, hybrid systems—such as variable-speed control pumped storage hydropower combined with volatile renewable energy sources (e.g., solar, wind) —reflect contemporary efforts to integrate hydropower into flexible and resilient energy systems.

Technological evolution has been accompanied by significant architectural transformation. Early hydropower installations were often housed in simple wooden enclosures or partially open structures, closely tied to the landscape and local construction traditions. As turbine sizes increased and operational requirements became more complex, fully enclosed powerhouses emerged, eventually leading to semi-submerged and cavern-type facilities that prioritize efficiency, safety, and environmental integration.

In recent decades, however, a renewed architectural trend—particularly evident in small hydropower plants—has sought to reopen these structures to the public realm. By exposing selected technological components beyond essential access points, contemporary hydropower facilities increasingly serve not only as energy-producing infrastructure but also as educational spaces. Through guided tours, training programs, and public engagement, these installations contribute to raising awareness of renewable energy, fostering technical literacy, and reinforcing the cultural and environmental value of hydropower in modern society (Fig. 2.1).

Harnessing the power of water: a historical journey of hydropower

The history of hydropower is a long and interconnected narrative that reflects humanity's enduring effort to harness natural forces for productive use. From simple mechanical water wheels to complex hydroelectric systems capable of supplying electricity to entire regions, hydropower has evolved alongside civilization itself. Spanning millennia and cultures, its development marks some of the most important milestones in technological, industrial, and energy history.

Ancient and preindustrial origins

The use of water as a source of power dates back to ancient civilizations, where flowing water was first harnessed to perform mechanical work. As early as 300 BC, water wheels were developed in regions including China, Greece, and the Roman Empire (Hansen

Figure 2.1 The Shimen pipeline small hydropower plant of the Shimen power plant (left and middle) and the Songlin unit of the Wanda power plant, Taiwan. Both examples represent the new trend in exposed technology solely for the site visits and educational (right) purposes. In these two cases a spiral casing is exposed to visually explain how water enters the turbine. Under normal circumstances the spiral casing is normally fully enclosed by concrete shells when in a vertical configuration, especially in Taiwan, due to vibrations caused by seismic activity. In the case of the Songlin unit, the pictured turbine is designed as a fully operable "educational/training" unit and is actually installed in parallel with a larger fully enclosed unit used primarily for energy generation.
Personal archive of Štefan Tkáč (left and middle 2025, right 2013)

Roger D.; Wikiwand). These early documented devices were primarily used for grinding grain, sawing timber, pumping water, and supporting agricultural and mining activities. In a slight variation, the wood-and-bamboo water wheels are still used in agriculture across Southeast and Far East Asia (e.g., along the Mekong River—China, Myanmar, Laos, Thailand, Cambodia, and Vietnam, but there are also cases in Japan, Taiwan, and Korea still fulfilling their original function and also provide a glimpse into the past living in coherence with the water).

In Mesopotamia, particularly in Sumer and Babylonia, hydropower-supported irrigation systems were essential to sustaining agriculture between the Euphrates and Tigris rivers (e.g., Syrian Norias of Hama—a battery of scoop wheels used to supply the local aqueduct, but cases could still be found along the Euphrates River in Syria and Iraq).

On the other hand, China made particularly notable early advances during the Han Dynasty (202 BC–220 AD), where vertical-axis water wheels powered trip hammers used for hulling grain, crushing ore, and even early papermaking. These innovations represented an early form of mechanized production and were later adapted and refined in Western civilizations. In modern China, including Taiwan, water wheels are still used to pump water into irrigation canals that feed paddy fields.

In Ancient Greece and Rome, hydropower was further advanced in both practice and theory. Marcus Vitruvius Pollio, 80–70 BC—after 15 BC known as Vitruvius—documented hydraulic machines in *De architectura* (Chapters V-VII) (Morgan Morris, H.) while earlier engineers such as Ctesibius of Alexandria (c.300 BC–c. 222 BC) developed piston pumps capable of lifting and pressurizing water (Morgan Morris, H.). Another enduring, improved invention by Archimedes of Syracuse (c.287–c. 212 BC), the Archimedes screw, with some sources dating it back to 300 BC, was originally designed for irrigation and drainage, remains relevant today, and has found a renewed application in ultralow-head hydropower systems due to its robustness and adaptability.

Across these early societies, water-powered machines enabled more efficient and consistent performance of repetitive labor. Although not yet used for electricity generation, these systems established the fundamental principles of converting kinetic energy into useful mechanical work, laying the groundwork for later industrial and technological revolutions.

Medieval developments in Europe

During the medieval period, hydropower became an integral part of Europe's economic and industrial life. Water wheels were widely adopted, powering mills for grain grinding, textile processing, metalworking, and woodworking. Both horizontal and vertical water wheels were developed, including undershot, breastshot, and overshot configurations, each suited to different hydraulic conditions.

By the late Middle Ages, water-powered mills were central to local economies, often forming the nucleus of early industrial settlements. The refinement of mechanical transmission systems during this period significantly increased efficiency and reliability, setting the stage for the transition to industrial-scale production. Water canals, originally used to bring water to industrial zones, are still commonly found in the urban fabric today.

Renaissance and the Industrial Revolution

The Renaissance brought further refinements in water wheel design, informed by advances in science and engineering. These improvements reached a turning point during the Industrial Revolution of the 18th century, when hydropower became a critical driver of early factory systems. One of the most influential examples was Richard Arkwright's Cromford Mill, established in 1771 in

England, where waterpower was used to drive cotton-spinning machinery and inaugurate a new model of industrial organization.

At the same time, the development of water turbines marked a major technological leap. In 1827, French engineer Benoît Fourneyron introduced one of the first practical reaction turbines, capable of producing significantly more power than traditional water wheels. This was followed by James Francis's development of the Francis turbine in 1849, which remains the most widely used turbine type today. Later innovations included the Pelton impulse turbine, developed by Lester Allan Pelton in the 1870s, and the Kaplan turbine, introduced by Viktor Kaplan in 1913, which enabled efficient power generation across a wider range of flow conditions.

The emergence of hydroelectricity

The late 19th century marked a pivotal transformation in hydropower history: the shift from mechanical energy to electricity generation. Early experiments demonstrated that water turbines could drive dynamos to produce electric current. In 1880, a turbine-driven dynamo provided arc lighting in Grand Rapids, Michigan, followed shortly by similar installations at Niagara Falls. These early systems relied on direct current, which limited transmission distance.

The breakthrough came with the adoption of alternating current, allowing electricity to be transmitted over long distances. One of the first commercial alternating-current hydropower plants was commissioned in 1893 at the Redlands Power Plant in California, using Pelton wheels and a three-phase generator. Around the same time, Nikola Tesla and George Westinghouse developed the first large-scale hydroelectric system at Niagara Falls, completed in 1895, which demonstrated the feasibility of centralized electricity generation and long-distance transmission.

Twentieth-Century expansion and mega projects

The 20th century witnessed the rapid global expansion of hydropower, driven by growing electricity demand and advances in civil and electrical engineering. Large dams and reservoirs were constructed across North America, Europe, Asia, and Africa. Iconic projects such as the Hoover Dam (1936) in the United States, the Grand Coulee Dam (1942), and the Aswan High Dam in Egypt (1970) exemplified the scale and ambition of this era. After World War II, hydropower became a cornerstone of national electrification programs and economic development strategies.

By the early 21st century, projects such as China's Three Gorges Dam—completed in 2008, and Baihetan Hydropower Station-completed in 2022—illustrated both the immense power-generating potential of hydropower and the growing complexity of its environmental and social implications.

Environmental awareness and modern innovations

From the 1970s onward, increased awareness of ecological disruption, community displacement, and landscape transformation led to greater scrutiny of large dam projects. Environmental impact assessments became standard practice, and some proposed developments were modified or abandoned. For example, the Cheay Areng Dam in Cambodia, the HidroAysén project for the Patagonia dam in Chile, or the Wanli hydropower plant in Taiwan.

In response, modern hydropower has increasingly focused on sustainability and innovation. Run-of-the-river systems, which minimize reservoir size and ecological disturbance, have gained prominence, along with the rehabilitation of existing infrastructure. Advances in turbine efficiency, fish-friendly designs, and digital control systems have further improved performance. Additionally, renewed interest in tidal, wave, and ultralow-head technologies reflects ongoing efforts to expand hydropower's role within a diversified renewable energy portfolio.

Legacy

The history of hydropower is a testament to humanity's ability to adapt natural forces to changing social, economic, and technological needs. From ancient water wheels to contemporary hydroelectric plants, hydropower has continuously evolved while remaining rooted in principles established thousands of years ago. As the world confronts the challenges of climate change and sustainable development, this long history offers valuable lessons on innovation, resilience, and the necessity of balancing energy production with environmental and social responsibility.

History of water wheel design

The systematic development of water wheel design, and ultimately modern water turbines, began in earnest in the 18th century. A decisive milestone was the publication of Architecture hydraulique (1737–1753) by the French engineer Bernard Forest de Bélidor

(1698–1761). In this four-volume work, Bélidor moved beyond empirical craft knowledge and laid out hydraulic engineering principles grounded in mechanics. His comprehensive treatise covered harbors, maritime works, pumps, mills, and water wheels, establishing a scientific foundation for future advancements in hydropower technology (Fig. 2.2).

Further significant progress was achieved by John Smeaton (1724–1792), a Fellow of the Royal Society, who greatly improved the efficiency of water wheels through rigorous experimentation and analysis. Smeaton's work exemplifies what would today be described as an interdisciplinary approach, combining theoretical insight with practical engineering expertise. His Apparatus for Determining the Efficiency of Water Wheels (Horniblow 1988). Enabled systematic comparison of wheel types and operating conditions. Among his most influential contributions was the emphasis on bucket geometry, demonstrating that carefully shaped buckets could extract

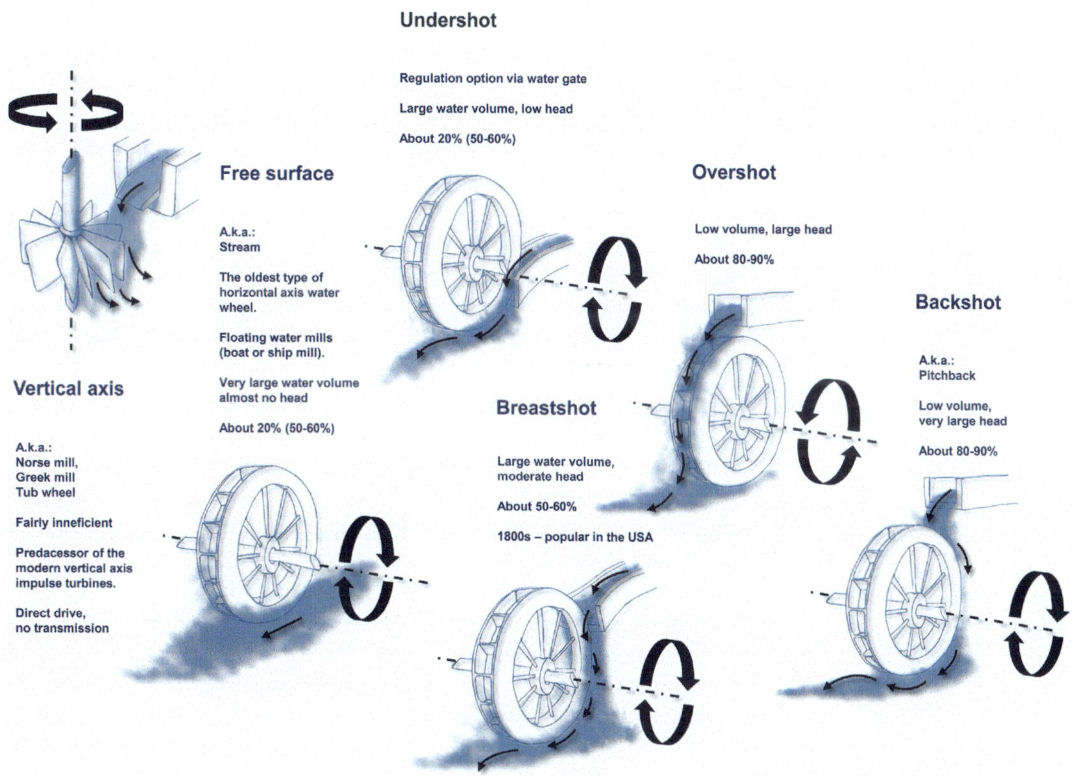

Figure 2.2 The general typology of the common water wheel configurations and their estimated efficiency.

substantially more energy from flowing water. These findings later informed the development of cross-flow turbines and impulse runners such as the Pelton wheel. For this reason, Smeaton's work is often regarded as marking the dawn of modern turbine development.

In the early 19th century, French engineer and mathematician Jean-Victor Poncelet (1788–1867) advanced water wheel design by applying mathematical principles derived from the work of Jean-Charles de Borda (1733–1799). Borda had linked water wheel efficiency to the relationship between water velocity before and after passing through the wheel. Building on this insight, Poncelet recognized that flat paddles caused unnecessary losses due to impact and splashing. His solution was the introduction of curved or cylindrical paddles that allowed a smoother transfer of kinetic energy. The resulting Poncelet wheel, constructed in 1838 and tested extensively using scaled models, represented a major leap in efficiency.

The first full-scale implementation of the Poncelet wheel occurred in the city of Metz, where it replaced an older installation and demonstrated an efficiency increase of approximately 33%. Owing to these impressive results, the Poncelet wheel rapidly spread throughout France and Germany, becoming a preferred upgrade for traditional water mills and their associated machinery.

Further refinements were introduced by engineers such as Alphonse Eléonor Sagebien (1807–1892) and Walter Zuppinger (1814–1889), whose designs continued to optimize blade geometry and flow conditions. The Poncelet, Sagebien, and Zuppinger wheels together represent a critical turning point in hydropower history. At this stage, wheel paddles evolved into carefully shaped blades or buckets designed according to hydrodynamic principles. Additionally, the need to guide and control the water stream led to the development of geometrically shaped chutes and flow-regulating mechanisms. These chutes were eventually enclosed to maintain water pressure, enabling operation under higher heads. From these innovations emerged concepts such as guide vanes in reaction turbines such as Francis and Kaplan, and adjustable distributors or nozzles in impulse turbines like Pelton and Turgo runners—clear descendants of the classical water wheel.

At the onset of the Industrial Revolution, these improved water wheels became essential prime movers for early industrial machinery. Extensive networks of industrial canals were constructed, serving both as transportation corridors and as controlled water supplies for wheels and early turbines. Many of these canal systems remain visible today within historic industrial urban landscapes.

Despite their advances in efficiency and scale, water wheels were gradually supplanted by compact and more versatile water turbines. The parallel rise of steam engines further accelerated this transition, as steam power offered greater flexibility, independence from watercourses, and reduced infrastructural requirements. Nevertheless, water wheels remained widely used in agricultural and forestry applications well into and beyond the Industrial Revolution.

In contemporary practice, water wheels continue to evolve, particularly in microhydropower and remote applications. They are frequently employed as flow-measuring devices, educational demonstrations, or low-impact energy converters in run-of-river systems. Additionally, modern water wheels play an important role in the adaptive reuse of historical sites, pumping water in irrigation canals, contributing to the preservation and promotion of industrial heritage while maintaining a tangible link between early hydropower technologies and their modern successors.

History of hydropower turbines

The development of hydraulic turbines represents a gradual transition from simple water-driven mechanisms to highly optimized machines designed for efficient energy conversion. Over time, turbines have been classified according to numerous physical and mechanical criteria; however, the most widely accepted classification is based on their working principles, particularly the interaction between pressure and velocity of the flowing water. According to the behavior of the water as it passes through the rotating machinery, hydraulic turbines are fundamentally divided into reaction and impulse types, with additional categories driven by specific medium properties.

Beyond this principal distinction, turbines may also be classified by changes in pressure and speed of the flowing medium (pressurized or constant-pressure operation), by the orientation of the runner axis (vertical, inclined or horizontal), by construction principles (open, partially enclosed, or fully enclosed designs), by the presence or absence of a spiral casing in reaction turbines, by adaptability to fluctuating flows through adjustable blades or wicket gates, and by size, as some turbines are specifically designed for small-scale applications or for powering a distributed power generation facilities.

Reaction turbines

The term *turbine* itself was introduced between 1822 and 1824 by the French engineer Claude Burdin (1788–1873), who derived it from the Greek word τύρβη (týrbe), meaning whirling or vortex. In his memoir Des turbines hydrauliques ou machines rotatoires à grande vitesse, presented to the Académie royale des sciences in Paris, Burdin laid the conceptual foundation for modern turbine theory. His work was soon followed by practical advances, most notably by Benoît Fourneyron (1802–1867), whose early reaction turbines demonstrated unprecedented efficiency and marked a turning point in hydraulic machinery.

Reaction turbines operate under pressurized conditions, where water flows continuously through the runner, and energy conversion occurs through both pressure and velocity changes. A key precursor to reaction turbine development was the Segner wheel, devised by Johann Andreas Segner (1704–1777), often considered one of the earliest true hydraulic turbines. Numerous interpretations of the Segner principle emerged, and it found widespread application in mills and early industrial machinery. One of the best-preserved examples is Barker's turbine, an improved Segner-type device powering a corn mill at the Museo Hacienda Buena Vista in Puerto Rico. Named after the British physician and inventor Dr. Robert Barker (–1745), this turbine is regarded as the only surviving pre-Scotch turbine and is often referred to in industrial archeology as Dr. Barker's water mill.

Further refinements of the Segner principle were introduced over time. Around 1775, Abbé Pupil modified the water entry from top to bottom, while James Whiteland introduced nozzle improvements in 1832. By 1839, James Whitlaw developed the Scotch Mill, also known as the Whitlaw or Scotch turbine, which became the most significant industrial application of the Segner-type turbine. These turbines were notably used on the Morris Canal in New Jersey to power inclined planes that hauled canal boats between different elevations. Today, the remains of the Morris Canal, listed on the National Register of Historic Places, are recognized by institutions such as the American Society of Mechanical Engineers as exceptional examples of hydraulic industrial heritage. The Segner turbine principle also influenced later technologies, including the development of modern sprinkler systems.

During the 19th century, reaction turbine design diversified further through the work of engineers such as Pierre-Lucien Fontaine (1809–1895) and Nicolas J. Jonval (also known as "Feu Jonval," 1804–1844). The Jonval turbine (invented in 1840), later refined by Clemens Henschel and André Koechlin, evolved into the

Henschel-Jonval and Jonval-Koechlin turbines, which were widely used in European industry. A major milestone was achieved by British-American civil engineer James Bicheno Francis (1815–1892), whose collaboration with American civil and mechanical engineer Uriah Atherton Boyden (1804–1879) led to the improvement of the Fourneyron-based outward-flow Boyden turbine in 1848, which became known as the famous Francis turbine. The Francis turbine combined high efficiency with broad applicability, and until today, it remains one of the most widely used hydropower turbine types worldwide.

The 20th century saw further specialization, particularly for low-head applications, with the invention of the Kaplan turbine by Austrian engineer Viktor Kaplan (1876–1934). Derived from the propeller turbine, the Kaplan design introduced adjustable runner blades, greatly enhancing efficiency under variable flow conditions. Numerous variants followed, including, for this time of bold technical solution, using a spiral casing without stay vanes, which became widely known in Europe as the Reiffenstein turbine, patented by Austrian engineer Dr. Ing. Manfred Reiffenstein (1894–1986), and since it was possible to combine it with Francis, Propeller, or Kaplan types, derivatives such as the Francis-Reiffenstein, Propeller-Reiffenstein, and Kaplan-Reiffenstein turbines were designed, often intended for small-scale installations. Other modern adaptations and developments followed. These include the Saxo turbine, Siphon, Semi-Kaplan designs with further fish-friendly adaptation, such as the modern progressive Alden turbine, along with mini systems like DIVE and Vortex turbines.

Impulse turbines

Impulse turbines, in contrast, operate under constant pressure conditions, where water jets transfer energy to the runner primarily through changes in velocity. Early examples include the Schwamkrug turbine developed by German designer Friedrich Wilhelm Schwamkrug (1808–1880) and designs by Louis Dominique Girard (1815–1871). The most influential impulse turbine, however, is the Pelton turbine, invented by Lester Allan Pelton (1829–1908), which became the standard solution for high-head, low-flow applications. Another significant contribution is the cross-flow turbine, independently developed by Hungarian engineer Donát Bánki (1859 -1922), German inventor Fritz Ossberger (1877–1947), and Australian engineer Anthony George Maldon Michell (1870–1959). Often referred to as the Banki-Ossberger-Michell turbine, this design is valued for its simplicity, robustness, and suitability for small hydropower projects

(Picone et al., 2025). The Turgo turbine further expanded the impulse turbine family by offering higher rotational speeds than the Pelton design.

Medium properties

In addition to classical reaction and impulse machines, a variety of turbines are driven by specific medium properties. These include the Tesla turbine, which relies on boundary-layer effects, as well as kinetic turbines, such as PROTUR (Precession Rolling Turbine), known as SETUR, and hydrokinetic turbines that extract energy directly from moving water in waves, rivers, or ocean currents without significant head. Such turbines are increasingly relevant in modern renewable energy systems, as evidenced by the biomimetic bioSTREAM tidal-current power conversion system developed by BioPower Systems. Another notable technology is the Archimedes Screw Turbine, a modern, reversed reinterpretation of the ancient screw pump, functioning as a low-head, static-pressure machine and conceptually related to the overshot waterwheel. However, because water's pressure and weight cause the screw to rotate and the system can be entirely enclosed, it is often referred to as a reaction turbine; therefore, it can belong to both.

Together, these developments illustrate how hydraulic turbines evolved through centuries of experimentation, engineering refinement, and adaptation to diverse environmental and industrial conditions. From early reaction devices to advanced modern turbines, the history of hydropower reflects both technological innovation and the enduring relationship between water, energy, and human ingenuity.

Conclusion

The history of hydropower is a testament to human ingenuity and adaptability, spanning from ancient civilizations to the present day. Early societies harnessed the mechanical power of water through simple devices such as water wheels, enabling more efficient agricultural and industrial processes. Over centuries, these technologies evolved—refined during the Renaissance and revolutionized during the Industrial era—culminating in the development of water turbines and the birth of hydroelectricity in the late 19th century.

Major engineering feats, such as the Hoover Dam and the Three Gorges Dam, illustrate hydropower's transformative impact on economies and societies worldwide. Yet, the expansion of hydropower has not been without controversy. Environmental and social concerns, particularly those arising from large dam projects, have prompted ongoing debates about sustainability and responsible development.

Modern innovations—including run-of-the-river systems, advanced turbine designs, and the exploration of tidal and wave energy—reflect a commitment to minimizing ecological disruption while maximizing renewable energy output. As hydropower continues to evolve, it remains a vital part of the global energy mix, offering valuable lessons in balancing technological progress with environmental stewardship and social responsibility.

In summary, the journey of hydropower is one of continuous adaptation and learning. Its rich history underscores the importance of integrating innovation with a deep respect for nature and communities, ensuring that future developments contribute positively to both energy needs and the well-being of our planet.

List of abbreviations

AST	Archimedes Screw Turbine
IEA	International Energy Agency
OTEC	Ocean Thermal Energy Conversion
PS + RES	Pumped Storage + Renewable Energy Sources
SETUR	Precession Rolling Turbine (PROTUR)
ULH	Ultralow Head
VSHP	Variable Speed Hydropower

References

Horniblow, J., (1988). Smeaton's apparatus for determining the efficiency of water wheels. https://collection.sciencemuseumgroup.org.uk/objects/co50744/smeatons-apparatus-for-determining-the-efficiency-of-water-wheels.

Hansen Roger, D., Water Wheels. https://www.waterhistory.org/histories/waterwheels/.

Morgan Morris, H.,. VITRUVIUS / THE TEN BOOKS ON ARCHITECTURE. https://www.gutenberg.org/files/20239/20239-h/20239-h.htm. Original by M. V. Pollio.

Picone, C., Sinagra, M., & Tucciarelli, T. (2025). A new regulation system for crossflow-type turbines for the water management industry. *Energy, 328*, 136695. https://doi.org/10.1016/j.energy.2025.136695.

Wikiwand. Wang Zhen (inventor). https://www.wikiwand.com/en/Wang_Zhen_(inventor).

3

Division

Introduction

Hydropower plants belong to the most versatile and adaptable civil engineering structures. Each hydropower plant has a base, mechanical, and electrical part, regardless of size. The base represents a specific civil engineering structure that responds to the preferences of the given environment, while the mechanical part and electrical parts are responsible for converting the potential and kinetic energy to electrical power. Over the years of practice, engineers have developed numerous configurations of these three features mentioned above. The given necessities of the case always set the division and hierarchy of the suitable application. This chapter summarizes common categories of division of hydropower plants.

Division, according to the arrangement of the powerhouse, usually follows the spatial requirements. The narrower, more unstable, and more prone to flooding a site, the more engineering solutions lean toward vertical arrangements. On the contrary, where access to technology is the priority, the horizontal arrangement is applied.

- **Horizontal**
  In the machinery's horizontal arrangement, the turbine's axis is horizontal or tilted (e.g., *bulb Kaplan turbine*). This arrangement allows relatively easy lateral maintenance procedures. All parts of the turbine are usually visible. The generator could be connected directly (*direct drive*) or via various transmission systems (e.g., *belt or gearbox*), leading to the option of the generator also being mounted perpendicularly to the main turbine axis (e.g., *Bevel gear*).
- **Vertical**
  In the vertical arrangement of the machinery, the axis of the turbine is strictly vertical (*e.g., vertical Francis turbine*). This arrangement requires axial maintenance procedures. The individual turbine and generator parts are visible only where needed via inspection openings. In modern designs, the generator is connected almost exclusively directly through coupling (*direct drive*). However, to enable easier generator maintenance and low axle speeds with high torque, some small hydropower plant cases

Hydropower Plants. DOI: https://doi.org/10.1016/B978-0-12-824520-0.00004-6

apply transmission (e.g., Bevel gear or, in the case of lower speeds, *even a belt*). In the vertical arrangements, a transmission could also be found in older hydropower plants or historical cases, mainly due to the same reasons.

Division according to the location and shape of the powerhouse is of utmost important as it directly reacts to the necessities of the given site. From the entire hydraulic structure of the hydropower plant, the powerhouse represents the most important and the most vulnerable part. It houses all the mechanical and electrical equipment along with the governing systems; thus, the location, size, and form (*shape*) strictly follow the hierarchy of importance. Generally, if the site and shape are priority measures for matching the safe and efficient power generation, we divide the hydropower plants into three basic forms. **Surface types**, **semiunderground types**, and the most complex and sophisticated **underground units**, also called the **cavern types**, as they are literally inserted into the natural or man-made caverns.

- **The surface types** are depicted as the oldest types of hydropower plants (Table 3.1). Their construction was initially based on a fairly narrow typology of machinery, which was originally predominantly all designed as horizontal due to weight, manual regulation, and necessary maintenance. The surface power houses are still very common solutions, especially for small hydro power plants, where the existing technology, despite being significantly modernized, reduced in size, and automated, is still available in horizontal axis arrangements. Due to its size, the surface type hydropower plants also represent the most adaptable and relatively affordable solution, which could be applied to all sorts of conditions, many times using only the very basic equipment; therefore, it is commonly seen as a sustainable energy solution to developing regions or remote regions with limited access (e.g., *UNDP's Rural Energy Development Programme (REDP) in Nepal or South Asia Subregional Economic Cooperation (SASEC) Project Off-Grid Component*).
- **Semiunderground types** represent an intermediate stage where the powerhouse civil structure had to be adapted to the evolving vertical axis machinery, as well as developing the safety of the powerhouse. In the semiunderground types, the water-related machinery (e.g., turbine, spiral casing, main valves, governor, draft tube) is located underground, while the electrical machinery (e.g., *generator, exciter, transformers, control room*) is located above the ground. This arrangement makes the power plant more resistant to natural disasters, especially potential flooding. The old semi-underground types had generators protruding into the generator room, requiring much higher ceilings to accommodate the

Table 3.1 The list of hydropower plants holding the world´s record.

Note	Name (year of completion)	Location	Installed capacity
• The **first private hydropower plant** in the world	William Armstrong´s Cragside (1878)	Rothbury, Northumberland, United Kingdom	4.5 kW
• The **first commercial hydropower plant** in the world	Vulcan Street Plant (1882)	Appleton, Wisconsin	12.5 kW
• The **largest hydropower plant** in the world • The **largest conventional hydropower plant** in the world	Three Gorges Dam (2008–2012)	Hubei, China	22.5 GW
• The world´s record holder in the **penstock head** (1883m) • The world's most powerful **Pelton turbine** (*423 MW per unit*) • The most powerful hydropower plant in Switzerland	Bieudron Hydroelectric Power Station (1993–1998)	Swiss Alps / Canton of Valais, Switzerland	1.269 GW
• The world's most powerful **Francis turbine** (*1000 MW per unit*)	Baihetan Hydropower Station (2017–2021)	Baihetan Dam in China	16 GW
• The world's most powerful **Kaplan turbine** (*204 MW per unit*)	Sinop Hydropower Station (2014–2019)	Mato Grosso, Central-West Brazil	0.408 GW
• The world's largest tidal power plant • 10x**Kaplan bulb turbine** (*25.4 MW per unit*)	Sihwa Lake Tidal Power Station (2003–2011)	Ansan City, Gyeonggi-do, South Korea	0.254 GW
• The world's first seawater pumped storage hydropower plant *(single reversible Francis turbine)*	Okinawa Yanbaru Seawater Pumped Storage Power Station (1987–1999) *Decomissioned in 2016*	Kunigami, Okinawa Japan	0.03 GW

(*Continued*)

Table 3.1 (Continued)

Note	Name (year of completion)	Location	Installed capacity
• The world's most powerful **Pumped Storage Hydropower plant** (*Storage capacity 40 GWh, Reversible Francis turbines)*	Fengning Pumped Storage Power Station (2013–2021)	Hebei Province, China	3.6 GW
• The world's most powerful **Run-of River Hydropower plant** (*Bulb turbines)*	Jirau Dam Hydropower plant (2008–2016)	Rondônia, Brazil	3.75 GW
• The world's **highest-elevation** hydropower plant (*5000 m above sea level*)	The Maerdang hydropower station (2011–2024)	Qinghai Province, China	2.2 GW
• The world's **largest** hydropower plant using water from the WWTP	North Head Sewage Treatment Plant (2010)	NSW, Australia	4.5 MW
• The **largest hydropower cascade** in the world (*12 dams, built between 1970 and 2021*) (Huang & Li, 2024)	1. Gezhouba dam (1970–1988 / 2715 MW) 2. Three Gorges Dam (1994–2003 / 22,500 MW) 3. Xiangjiaba dam (2006–2012 / 6000 MW) 4. Xiluodu dam (2005–2013 / 12,600 MW) 5. Baihetan dam (2017–2021 / 16,000 MW) 6. Wudongde dam (2015–2021 / 10,200 MW) 7. Guanyinyan dam (2008–2014 / 3000 MW) 8. Ludila dam (2007–2013 / 2160 MW) 9. Longkaikou dam (2007–2013 / 1800 MW) 10. Jin'anqiao dam (2003–2010 / 2400 MW) 11. Ahai dam (2008–2012 / 6000 MW) 12. Liyuan dam (2008–2014 / 2400 MW)	The Yangtze River, China	87.7 GW (*Combined*)

maintenance crane. Modern types have the generator encapsulated for better cooling and fire protection via a fire-extinguishing atmosphere (generally *based on carbon dioxide or nitrogen gas*). The generator enclosure could be fully visible in the generator hall or embedded in the vertical construction as part of the substructure. In the latter case, only the generator´s cover, sometimes lit with the extended exciter, is visible.

- Ironically, **underground units or cavern types**, despite being pictured as the most advanced hydropower engineering solutions, were applied quite early, utilizing various underground structures such as caves or mining shafts. The oldest underground hydropower plant is the Snoqualmie Falls Cavity Generating Station (American Society of Civil Engineers, 2025), built between 1898 and 1899 in King County, Washington, United States. The machinery room lies 250 ft (76.2 m) below the surface, and the power plant is fully functional even today.

The underground units proved to be resilient to earthquakes as the main structure is embedded in the rock stratum, hence adopting the same frequency as the oscillating mountain. The structure is also shielded against direct sabotage attempts or attacks, and, to a certain extent, the cavern provides protection against floods and typhoons. However, specific security measures must be implemented (e.g., *watertight blast door*).

In most cases, underground units use a vertical-axis arrangement, even though Pelton turbines are commonly designed as horizontal. However, there are cases where either Francis or Pelton turbines in horizontal-axis arrangements are installed in caverns. This happens because of the cavern typology governed by the geotechnical conditions of the site, accessibility, (installed capacity), or simply the size, thus the construction budget of the respective unit.

Due to the relatively strong security of underground units, the installed capacity is usually very high; in modern units, the overall installed capacity almost always exceeds 1000 MW. The largest underground unit is in Quebec, Canada. It is called the Robert-Bourassa generating station, and it has 5616 MW installed in 16 vertical-axis Francis turbines (*each turbine 351 MW*). The annual generation equals 31,000 TWh.

Some units also use reversible turbines, making them an underground pumped-storage hydropower plant. The largest underground pumped-storage hydropower plant, with a total installed capacity of 3600 MW, divided into 12 vertical-axis reversible Francis turbines (each with 300 MW), was commissioned in 2021 as the Fengning Pumped Storage Power Station in Hebei Province, China. The annual generation equals 34,200 GWh.

Therefore, the underground hydropower plants are a vital energy system on a regional or even national scale.

Cavern typology

Choosing the right shape for the underground powerhouse cavern is subject to multiple stress situations, and the final overall position of the cavern is determined by the geometry of structural discontinuities in the rock mass. In an ideal scenario, the shape itself is a direct response to the distribution, size, and vectors of the external forces acting on the cavern´s crown and side walls. The typology of the artificial underground caverns is thus quite wide, hence it is hard to specify a precise shape as it could range from a simple rectangular cross-section with reinforced corners up to fully elliptical or circular. However, an artificial underground cavern designed for a hydropower plant has, in most cases, a vertically elongated cross-section due to vertical axis turbine arrangements, which is the most typical, also due to the possibility of additional horizontal stabilization of the cavern walls. Generally, there are recognized further 3 shapes and their geometric modifications.

- **Mushroom-shaped cavern with a concrete roof arch** is commonly used to bear load mainly vertically. The mushroom itself allows additional reinforcement and roof support structures to be applied. Also, the edges of the mushroom allow for ideal installation of the maintenance crane rails. There is no fixed geometry of the mushroom roof, as it varies from case to case. Lateral reinforcement is required for the walls.
- **Horseshoe-shaped cavern** is applied in a less hard rock strata where forces acting on the roof may not be exclusively or predominantly vertical. Just like in the case of the mushroom type, the walls had to be additionally reinforced for lateral forces. The maintenance crane rail is usually embedded into the side wall structure or has a structure of its own.
- **Elliptical cavern** is the most complicated for construction as it has an elliptical wall, which are prestressed to sustain the lateral forces. This solution is usually applied in soft rock strata or unstable regions challenged by potential earthquakes. The elliptical cavern is designed to be fully self-supported while resisting forces from all potential directions with relatively small additional reinforcement. The cavern itself represents one of the advanced and sophisticated underground structures; it is also the most expensive solution, which is the very reason why it is not applied so often as the additionally reinforced Horseshoe or Mushroom type.

Division, according to the power medium

Generally, hydropower plants are associated with a river water being the medium for generating electricity via gen-set technology; however, in recent years, technological development and making the environmental laws stricter, along with the rising need for efficiency and energy, demands the hydropower plants to adapt to use water in other than clean forms.

Fresh drinking water, whether originating from the reservoir or artesian wells, is often transported to the city via aqueducts or pipelines due to high elevation; hence, the pipelines had to be interrupted to relieve pressure, which is generally done by relief valves. These valves could be commonly replaced by a Pelton runner (Francis and other pipeline turbines are applied as well) to harvest that surplus energy. The technological pressure for this scenario is put on the sealing and lubrication systems as well as the turbine material itself, which must be perfectly clean. Using the power of fresh drinking water is not a new technology and has been developing with the modern urban delivery piping system. However, with the developing precision of the turbines, efficiency and adaptability have increased, and thus many cities already implement it. Still improving turbine technology that allows for applications in the drinking water supply system could be found in two general configurations. There is an application before and after the purification facility, even though in both cases the technology must be clean and with the ability to be directly bypassed due to potential revision and maintenance stops, the stricter regulations are placed on the technology in the second configuration. Some of the cases are listed in Table 3.2; a specific city example could be found in Athens (4600 kW), Hong Kong (500 kW), Sandiego (4500 kW), and there are multiple small, mini, micro, and pico applications placed directly in the purification plants across the world. An interesting network of these direct installations could be found across Japan.

Figs. 3.1 and 3.2 show configurations for hydropower plant tap water applications. Two specific examples from Japan, namely Yokohama and Fukushima, demonstrate the installation of a hydropower turbine before and after the purification process.

Industrial and agricultural water, although many times not different from river water, refers to either harvesting water energy from the industrial water outlets, mostly referring to cooling cycles for factories or even thermal or nuclear power plants, and water diverted for agricultural purposes, such as field watering or as a supply for the paddy fields, along with the fish farms. As the agricultural water is mostly transported via dedicated water canals,

Table 3.2 The list of some of the hydropower plants using drinking water.

Note	Name (year of completion)	Location	Installed capacity
• Historical hydropower plant, using water from the Yanming mountain artesian wells. Water passes the horizontal axis dual jet Pelton runner. The power plant was decommissioned in the '70s	Sanjiaopu Hydropower Plant (1930)	Taipei, Taiwan	500 kW
• The only currently operable hydropower plant using drinking water in Taiwan. Water passes through a horizontal Francis turbine.	Lijia Hydropower Plant (2023)	Taitung, Taiwan	130 kW
• Direct pipeline installation of a battery of hydropower units.	The Lucid Energy LucidPipe System (2014)	Portland, Oregon, US	200 kW
• Using water from the drinking water distributing network	Ain Leghwaibe	Ain Leghwaibe, Northeast of Lebanon	743 kW
• Using water from the drinking water reservoir	Poggio Cuculo Hydropower plant (2010)	Arezzo, Italy	44 kW
• Using water from artesian wells, using a horizontal axis dual jet Pelton runner	Balme Hydropower Plant (2008)	Balme, Piedmont, Italy	1434 kW
• Using water from the drinking water distributing network	Troistorrents Hydropower Plant (1999)	Troistorrents, Switzerland	70 kW

• Using drinking water from the glacier to power up the vertical axis Pelton runner	La Zour Hydropower Plant (2004)	Savièse, Switzerland	465 kW
• Using water from the drinking water distributing network	Saint Jean Hydropower Plant (2009)	Saint Jean, Switzerland	102 kW
• Using two twin-jet Pelton turbines and 445 m gross head. Water is coming 1.6 m long from a spring collection tunnel. Mühlau is one of the largest hydropower plants based on drinking water.	Mühlau Hydropower Plant (1952)	Innsbruck, Austria	5750 kW
• The water originated in the Alps mountains and later passed horizontal Francis turbine.	Vienna Mauer Hydropower Plant (2006)	Vienna, Austria	500 kW

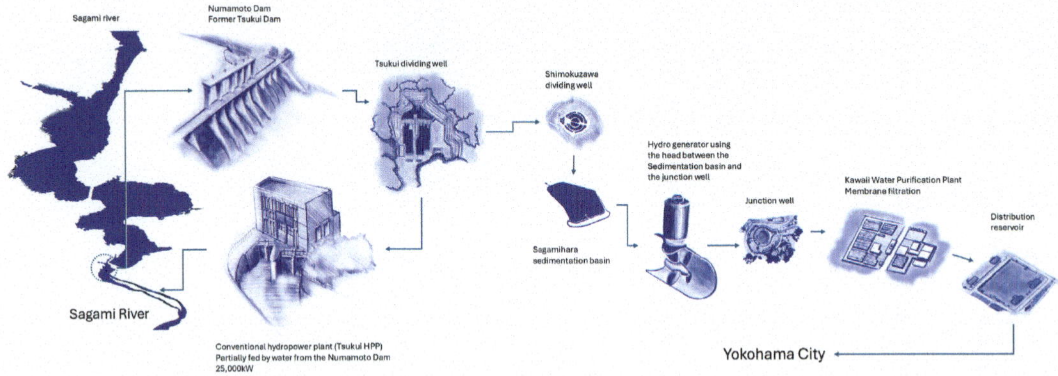

Figure 3.1 Configuration with turbine placed before the purification plant, example of Kawai Water Purification Plant, Kamikawaicho, Asahi Ward, Yokohama, Kanagawa, Japan.

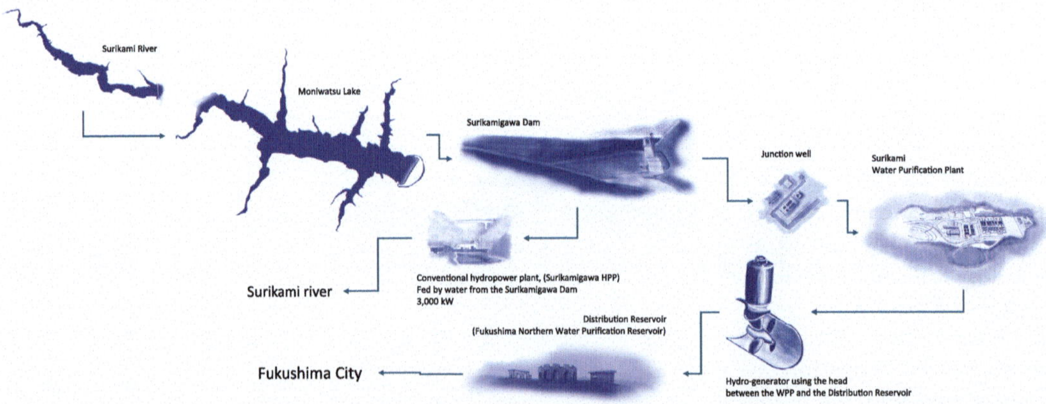

Figure 3.2 Configuration with turbine placed after the purification plant, example of Surikami Water Purification Plant, Iizaka, Fukushima, Japan.

the power plants suitable for this scenario are called canal power plants and are usually designed for low head or even ultralow head with turbines in horizontal arrangement. In both cases, water might require additional water treatment.

Wastewater from a WWTP is one example. In this scenario, the powerhouse is typically placed after the WWTP, avoiding the use of the untreated raw sewage water, which is beside solid waste, also possibly mixed with all sorts of acids, salts, alkalis, and hydrogen sulfide (H2S), all potentially interacting with the material of the turbine's blades, leading to corrosion and imbalance. However, using basic screens, sedimentation basins, along with manufacturing turbines from hard materials like stainless or duplex steel, with a combination of ceramic coating solution, allows the installation of the turbine even before the WWTP. These technical solutions allow the use of even raw, untreated water, which would normally lead to the quick wearing off of turbines to generate electricity. A typical technical solution would be a Kaplan or Cross Flow turbine, both with significantly enlarged gaps between the blades, usually angled, and with a special surface texture to minimize sedimentation.

Salty water is another example where turbine technology needs to be adapted, mainly due to potential corrosion. Using salty water brings several advantages, one being the fact that salty water is denser, hence the system could be smaller and more effective. Except for the Tidal Power Plant (TPP), the salty water is almost exclusively used in the case of Pumped Storage Hydropower Plants (PSHPP) built close to the seashore. However, there are very few examples of this arrangement; TPP is commonly used as a viable reference.

Fluid with higher density also represents a PSHPP solution where the water is replaced with a fluid denser than water, and just like in the case of salty water, it is a minerals-based fluid, and it allows for the technology to be simply smaller and more efficient. However, in contrast to the salty water scenario, the use of a denser fluid system cannot be designed as an open-loop, thus it is a closed facility constantly recycling the same fluid between the upper and lower reservoir (Informa, 2022).

Division according to the arrangement of the civil hydraulic system

The hydropower plants and their integral parts form a unique merger of civil and mechanical structures, resulting in a civil hydraulic system tailored to the given location. Based on the arrangement of the essential parts of this civil hydraulic system, the power plants could be divided into conventional, run-of-river, and pumped-storage hydropower plants.

Conventional hydropower system

A significant portion of conventional hydropower systems is usually formed by an impoundment structure that stores water. These structures are mostly represented by a large, artificial reservoir or dams, which are also the main cause for the ecological challenges and mostly shade the positive outcomes that the conventional hydropower system represents. Regarding the further division, the conventional hydropower systems work on the scale principle between the given head and available (flow volume) discharge.

- **Ultralow head (ULH)** conventional hydropower systems typically work with 3 m head and lower, which can sometimes drop as low as 0.5 m. In most cases, installed capacity does not exceed 1 MW.
- The ULHs can be set as surface or semiunderground units, but since their installed capacity typically refers to pico, mini, and micro systems, many cases involve canal types with stable, regulated discharge. The installed turbines are typically Kaplan, propeller, or Archimedes-screw-based systems, but vertical crossflow designs (JAG Seabell´s ULH STREAM; Hydro power, 2020) are also used in some scenarios. Some turbines in this category are retractable and nonpermanent; hence, they can be used to measure flow parameters or serve as an experimental turbine testing platform, e.g., the Gorlov helical turbine (GHT). The generators are usually connected via belt, chain, or are directly driven.
- **Low-head conventional hydropower systems use more kinetic energy of the water rather than the potential energy**, as the main parameter is the water discharge. Head, on the other hand, is considered lower, hence the name. The typical range is 20–30 m with a surface or semisurface arrangement of the powerhouse. The turbines installed in these scenarios are reaction turbines with rather slow motion but high torque, allowing the installation of generators with larger capacity. The turbine axis, especially in river impoundment scenarios, is usually configured horizontally or slightly inclined, making it ideal for Kaplan or propeller-type turbines. However, if the available head generated by the dam or impoundment structure, in case of reservoir scenarios, permits it, vertical Francis, Kaplan, or even Cross-flow (or Banki-Michell) turbines could be applied. The Gen-sets are either designed as direct-drive or with speed-increasing drives, such as a parallel gearbox or a bevel drive. The belt transmission is also very typical in installed capacities below 150 kW. The direct drive is mostly used in vertical arrangements where the speed is adjusted by the turbine shape and regulating equipment (governor, wicket gates).
- **Medium-head conventional hydropower systems balance kinetic and potential energy**. Although the main parameter is the projected head, or potential energy, the discharge is equally important given the medium head range of 20–300 m. This is the most common small hydropower plant category, covering the most common head range and scenarios, the combination of turbine technology and generator connections is the widest, ranging from all sorts of axis arrangements, including inclined axis (e.g., inclined

bevel-driven Kaplan turbine) with direct drive, gearbox driven, or in lower installed capacities (e.g. up to 600 kW but typically 150 and below 100 kW), even belt transmission can be applied. Turbines in this category tend to be the most flexible and adaptable for a wide head and discharge range.

- **High-head conventional hydropower systems, on the contrary, benefit more from the water´s potential energy than kinetic energy**. This is ensured by a high-pressure steel tube called a penstock. The penstock ensures stable water pressure, which is later used by so-called impulse turbines, which are almost exclusively used for high-head, low-discharge scenarios. In a conventional hydropower plant, the most common turbines are either high-speed Francis or Pelton, available in horizontal or vertical configurations. In the case of Pelton, multijet and multi-runner variations are also available, depending on the given powerhouse space and water discharge. Pelton turbines operate at atmospheric pressure; they use high-pressure water, and the runner is turning at a high speed; therefore, they are commonly designed as direct-drive turbines. In the case of smaller systems like further run-of-river (ROR), the variety of turbines and configurations significantly expands, for example, Turgo, or the mentioned high-speed "Express" Francis turbines. The typical range of high-head turbines starts at 300 m. High-head systems are also common for pumped-storage hydropower plants (ranging between 100 and 1400–1500 m in multistage configuration).

The run-of-river (ROR)

The run-of-river (ROR) hydropower systems represent a type of renewable energy generation that uses the river's natural flow (only a portion of the natural flow) to produce electricity without constructing large reservoirs or dams. Despite ROR often being connected with the term small hydropower, they represent different hydropower divisions. The installed capacity of a hydropower plant could refer to a small hydropower category, but it still does not fulfill the requirements for an ROR and vice versa. For example, the division of the power plants by the installed capacity for small hydropower plants varies from country to country.

On the other hand, rather than dwelling solely on the installed capacity, the ROR systems mainly focus on the form of harnessing the natural stream´s energy to drive turbines and convert water flow into power. While there is no strict regulation on retention capacity in the case of small hydropower plants, there are quite firm boundaries on ROR retention systems. For example, not changing the natural course of the river. Therefore, there are several variations of ROR systems, each suited to serve the various economic models with a reflection of the different river profile conditions:

- **Submerged or floating (hydrokinetic) ROR systems** use underwater turbines or are mounted on floating platforms to capture the river's kinetic energy directly. They can be constructed as fixed/permanent or portable. The durability and installed capacity are usually the downsides, while the flexibility, size, and, often, the bureaucracy are the upsides, which are why these ROR systems are seen as temporary testing platforms to harness flow energy data or test a new type of turbine.
- **Diversion weir ROR systems** use a small dam-like structure that diverts a portion of the river's water flow to a power station, which is used to generate electricity. These retention structures are designed to respond to fluctuations in the river rather than storing water for extended periods. In many cases, the flexibility of the impoundment structure is the main reason it is installed, e.g., the flexibility of an inflatable dam against floods.
- **Pondage ROR systems** involve creating a small storage area, or pond, where water is temporarily stored to ensure a steady flow to the turbines, even when river conditions fluctuate. Compared with the Diversion weir scenario, which is usually applied in rivers, pondage ROR systems are used in high-mountain areas, mainly to handle seasonal flow changes.
- **Dam-Toe ROR systems** represent a scenario where turbines are installed near the base of an existing dam to take advantage of the existing water flow without requiring new large-scale infrastructure. These are very common systems in hydropower refurbishment, dam restoration, or dam conversion to a hydroelectric structure.

In general, in the recent hydropower industry, ROR systems are seen as environmentally friendly due to their relatively minimal impact on river ecosystems. Their flexibility and scale offer an efficient and sustainable way to generate power from flowing water, whether as part of the grid infrastructure or as a stand-alone solution for decentralized generation.

PSHPP—pumped storage hydropower system

Pumped storage hydropower (PSHPP) is a type of hydropower technology designed to store energy in the form of Newtonian potential, broadly known as gravitational potential energy. This potential is achieved by using two reservoirs at different elevations. PSHPP is a form of energy storage that can help balance electricity supply and demand, enhance grid stability, and support the integration of renewable energy. In 2024, the PSHPP installed capacity globally totaled 200 GW, representing 94% of the world's long-term energy storage capacity. Since the water or liquid medium can be used repeatedly, the PSHPP is the largest rechargeable battery in the world (International Hydropower Association, 2022). The working principle of PSHPP is to pump water from a lower reservoir to an upper reservoir during periods of low electricity demand (when excess energy is available) and release the water back to the lower reservoir through turbines to generate electricity during peak demand periods. In general, there are two types of PSHPP. The open and closed-loop systems. The open-loop or natural flow-integrated systems are, in a sense, naturally fed by existing inflow, and the closed-loop systems are based on a stand-alone principle that is not connected to any of the outer natural flows.

Typologies of pumped storage hydropower

Closed-loop system
Hydrologically speaking, it is a stand-alone system that can be designed in a surface or subsurface configuration. While in the Surface configuration, the reservoirs are connected through a closed loop of pipes and tunnels that avoid the use of natural water sources, in the Subsurface configuration, both reservoirs, along with the powerhouse and connecting piping, are located underground, minimizing the land footprint and reducing the environmental impact.

Open-loop system
In contrast to the Closed-loop system, the Open-loop system is connected to the natural flow, which feeds either the upper or lower reservoir. The system could be equally designed in surface and subsurface configurations, where, in the case of surface one, it uses locally available natural water resources such as rivers or streams to feed natural or artificial lakes serving as upper or lower reservoirs. In the Subsurface configuration, the reservoirs are fed by natural underground aquifers or subsurface water bodies.

In the open-loop system, based on the configuration toward the connection/access to the natural flow, there are 3 working scenarios of the PSHPP.

Natural feed to lower reservoir represents the most common setup, where the lower reservoir receives water from natural sources (e.g., rivers) combined with a conventional hydropower system. The upper reservoir is usually man-made and located above.

Natural feed to the upper reservoir enhances the energy generation capacity of a hydropower cascade, where natural water sources

replenish the upper reservoir. In an ideal scenario, this working scheme would use natural river inflow to an artificial upper reservoir, but an artificial diversion structure from another riverbed to feed the upper level is also a common solution.
Natural feed to both reservoirs is considered the ideal or almost idealistic scenario. In this scenario, the upper and lower reservoirs receive water from natural sources, providing maximum efficiency and sustainability. On the other hand, this would mean that the PSHPP is part of the existing hydraulic cascade or at least has great potential to develop one.

The oldest PSHPP was built in 1907 in Switzerland, and in the 1930s, a reversible hydroelectric turbine became available. Since then, the PSHPP technology has significantly expanded and remains a target for ongoing improvement and development. Some of the current technological solutions are:

- **Freshwater coastal reservoirs**. These systems use freshwater from river floods stored in coastal reservoirs. Water is pumped from the lower reservoir to the upper reservoir, and energy is generated when water is released back.
- **(HD hydro) high-density pumped hydro**. This technology was developed by the UK-based firm called RheEnergise. It is a closed-loop system using fine-milled mineral-based fluid named R19, which is about 2.5 times denser than water, to increase energy storage capacity. Due to denser fluid, the entire system can be smaller, and using high-efficiency turbines and pumps allows for more compact installations with higher power output. Hence, the cost for the civil structures, which represents a significant portion of each power plant, could be significantly reduced while keeping the same or even higher output as the original scale facility using freshwater (Informa, 2022).
- **Use of seawater**. Some coastal systems are designed to use seawater instead of freshwater. Despite the potential saltwater corrosion and barnacle growth, seawater-pumped storage systems offer significant potential for large-scale energy storage. Between 1987 and 1999, the Okinawa Yanbaru Seawater Pumped Storage Power Station in Okinawa, Japan, was built as the world´s first seawater PSHPP in the standard arrangement with a top artificial lake and sea acting as a bottom reservoir. The power plant was decommissioned in 2016 due to profitability; however, it was a proof of concept that has recently been followed by Lanai Pumped Storage in Hawaii, Spencer Gulf Project in Australia, and even the San Onofre project in San Diego County, California, which intends to repurpose the stranded infrastructure of the decommissioned local nuclear power plant.

- **Underwater reservoirs**. Underwater reservoirs, such as those located on the sea floor, store energy by pumping water into submerged caverns. This method minimizes land use and reduces environmental impact.
- **Underground cavern reservoirs**. In this technology, reservoirs are located deep underground within natural caverns or man-made tunnels, providing a compact and efficient solution for pumped storage.
- **Variable speed pumped storage power plant (VSPSPP).** Compared to the traditional fixed speed pumping gen-set technology, the VSPSPP allows flexibility of the rotation speeds, making them more resilient to potential adjustments to fluctuating outputs from the wind or solar farms. VSPSPPs are also more efficient and reliable due to their variable-speed nature, and they can support grid regulation via frequency or voltage regulation. On the other hand, these systems are also relatively new and thus require higher initial costs and additional cooling technology. However, in contrast to the mentioned fixed speed VSPSPPs, the advantages are often questioned for potential retrofitting.
- **Large- or small-scale decentralized systems in conventional or hybrid configuration**. Just like any other technology, PSHPPs can be designed at small, medium, or large scales. However, on a small scale, they are an ideal local energy storage solution that could be scaled down to microsystems supporting either a larger manufacturing industry or a town, and, when designed as a hybrid PSHPP, could significantly contribute to the development of energy 0 or even energy-positive communities.
- **Ternary sets (FSPSPP/VSPSPP)** are the most modern pumped storage technologies designed for rapid, almost uninterrupted switching between pumping and generation modes, which is vital for stabilizing the RES-reach energy mix grids. The composition can be vertical or horizontal and consists of three main components connected by a single shaft: turbine (usually a Francis or Pelton turbine), motor-generator, and a separable (usually via torque converter) multistage pump. All components rotate in the same direction; this arrangement allows the complete elimination of sleeve/pony engines. It is also the most expensive solution and requires the most space.

PSHPP is a crucial technology for supporting the transition to renewable energy, as it enables the storage of energy from variable sources like solar and wind and provides a reliable, on-demand power supply when needed. What is more, as VSPSPP technology advances, hybrid solutions, even when paired with volatile wind or solar farms, can significantly contribute to grid stabilization and resiliency.

Others

Despite being the oldest renewable energy industry, the hydropower sector is continually evolving as new technologies and forms are developed to harness the power of moving water. Among these other forms of hydropower are wave and ocean current energy harvesting, as well as hybrid solutions, in addition to conventional dam-based systems, RORs, and PSHPPs.

- **Hybrid solutions** represent a combination of two or more renewable energy technologies, like wind, solar, and various water-based systems, to create a more reliable and efficient system. The hybrid solution offers even greater flexibility and robustness against natural resources that are prone to fluctuations due to long-term seasonal changes or more immediate weather shifts. (e.g., Solar-powered pumped storage hydropower plant or a pico hydropower system reinforced by a solar unit with the battery pack to store excessive energy)
- **Wave energy harvesting** captures the kinetic energy of ocean waves, transforming it into electricity using floating or submerged devices that move with the water's motion. There are various technologies and devices on the market, all designed to capture wave motion. The commonly recognized divisions are open-water solutions, such as Point Absorbers (e.g., active buoy) or Overtopping Devices, and shore-connected solutions, such as Oscillating Water Columns (OWCs) or anchored Attenuators (e.g., Pelamis).
- **Ocean current energy harvesting** taps into the steady flow of ocean currents, using underwater turbines or an inverted pendulum to convert their kinetic energy into power, much like horizontal-axis wind turbines (HAWT), vertical-axis wind turbines (VAWT), or recently introduced vortex-induced vibration bladeless oscillator does for the air.

These diverse hydropower solutions offer exciting possibilities for clean, renewable energy, each addressing different environmental conditions and energy needs.

Conclusion

This chapter has shown that hydropower facilities are not all the same, nor do they fall into a single category; instead, they represent a wide variety of engineering techniques determined by site, water characteristics, and energy needs. This chapter has divided hydropower facilities by divisions based upon how the machines are arranged, the location of the powerhouse on the site, the type of power medium used, and the construction of the civil–hydraulic systems. All divisions show that adaptability is the primary strength of hydropower. The equipment can be arranged horizontally or vertically; the powerhouse location can be on the surface, partially underground, or in a hillside cavern. In all cases, careful consideration was given to the elements of space availability, operational efficiency, safety, and longevity in the design of the facilities.

The comparison between horizontal and vertical layouts, as well as surface, semiunderground, and fully underground powerhouses, highlights how design choices are shaped by spatial limits, safety requirements, maintenance demands, and long-term durability. At the same time, the growing use of nontraditional water sources—such as drinking water networks, industrial flows, wastewater, seawater, or high-density fluids—demonstrates how hydropower technology continues to adapt to environmental regulations and efficiency demands.

By examining conventional schemes, run-of-river systems, and pumped storage hydropower plants, the chapter also illustrates how hydropower can serve very different roles, from localized energy production to large-scale grid stabilization. In particular, pumped storage stands out as a key technology for the future, providing energy storage and flexibility essential for integrating wind and solar power.

Overall, the divisions presented in this chapter provide a clear framework for understanding hydropower as a highly adaptable and evolving energy system—one that bridges historical engineering principles with modern technological innovation and remains central to sustainable energy strategies worldwide.

List of abbreviations

CO2	Carbon dioxide
FSPSPP	Fixed speed pumped storage power plant
GHT	Gorlov helical turbine
GWh	Gigawatt-hour
HAWT	Horizontal-axis wind turbine
HD Hydro	High-density pumped hydro
JAG	Japan Asia Group
kW	Kilowatt
MW	Megawatt
OWC	Oscillating water column
PSHPP	Pumped storage hydropower plant
REDP	Rural Energy Development Programme
RES	Renewable energy sources
ROR	Run-of-river
SASEC	South Asia Subregional Economic Cooperation
TPP	Tidal power plant
ULH	Ultralow head
VSPSPP	Variable speed pumped storage power plant
WWTP/s	Wastewater treatment plant/s

References

American Society of Civil Engineers. (2025). Snoqualmie Falls Cavity Generating Station. Retrieved February 2, 2025, from https://www.asce.org/about-civil-engineering/history-and-heritage/historic-landmarks/snoqualmie-falls-cavity-generating-station.

Huang, Z., & Li, H. (2024). Dams trigger exponential population declines of migratory fish. *Science Advances, 10*(19). https://doi.org/10.1126/sciadv.adi6580.

Hydro power. (2020). Micro hydropower system (ultra-low head) | Climate Technology Centre & Network | Tue, 05/19/2020. Retrieved February 2, 2025, from https://www.ctc-n.org/products/hydro-power-micro-hydropower-system-ultra-low-head.

Informa. (2022). A new generation of pumped hydro: RheEnergise's high-density hydro® solution opens up geographies. Retrieved February 2, 2025, from https://www.current-news.co.uk/a-new-generation-of-pumped-hydro-rheenergises-high-density-hydro-solution-opens-up-geographies/.

International Hydropower Association. (2022). Pumped storage hydropower: Water batteries for solar and wind power. Retrieved February 2, 2025, from https://www.hydropower.org/factsheets/pumped-storage#:~:text=Pumped%20storage%20hydropower%20is%20the,it%20a%20rechargeable%20water%20battery.

4

Current technology

Introduction

Hydropower technology, despite being the oldest renewable energy harvesting technology on the market, is still evolving, especially in terms of adaptability and efficiency. With the development of still more precise manufacturing and material solutions (e.g., coatings), hydropower turbines could be designed and manufactured with stunning precision from a material (coated or plain) in direct response to the working environment, meaning the water property. Beyond standard river water conditions, modern hydropower turbines can operate in environments containing high concentrations of minerals, acids, or bases, enabling energy harvesting even from water systems with high risk or strict environmental restrictions, such as water treatment plants, tap water pipelines, and wastewater treatment plants, especially when installed before the treatment process. These advanced technologies allow countries with almost fully utilized feasible hydropower potential to focus on more targeted solutions, such as energy recovery, retrofitting old units, supporting decentralization or grid robustness, and building and improving pumped-storage energy technology to shorten the response time to grid fluctuations and to more efficiently back up the generated power.

With the advance of digital technologies, hydropower plants are not falling behind, and so-called Smart Hydropower Systems are already available, especially in the small hydropower market. Such innovative technologies usually incorporate the IoT (Internet of Things), specifically sensors and devices that monitor parameters such as water flow, turbine performance, and environmental conditions, enabling real-time data analysis and decision-making, as well as predictive maintenance. Advanced analytics and machine learning algorithms help predict equipment failures before they occur, reducing downtime and maintenance costs.

The hydropower turbine technology could thus be divided according to the scale and purpose into three categories, namely the technology for the large scale systems where size, installed capacity or retrofitting are the key parameters, the pumped storage technology where efficiency achieved via variable-speed control better interact with the grid inbalances usually caused by the energy

Hydropower Plants. DOI: https://doi.org/10.1016/B978-0-12-824520-0.00003-4

mix that consists of multiple volatile renewable energy technologies, and last but not least the small scale technology focused on environment needs (fish friendliness), robustness, versatility and efficiency.

Large scale

The large-scale turbine industry still operates with Impulse and Reaction turbines, reflecting the head. Although the range of modern reaction turbines, such as Francis-type turbines, spans from 1 to 600 or even 900 m, generally, when exceeding 100 m, impulse turbines are considered, as they operate on water pressure rather than flow discharge; thus, the higher the head, the better the performance. For comparison, the operating range of the modern Pelton type is 50 to 1800 m. The large scale Alden turbine technology is recently in consideration.

Impulse turbines

Impulse turbines, which refer to hydropower runners, typically use specially designed blades/buckets to capture the impulse generated by a single or multiple water jets. Regarding energy conversion, Impulse turbines capture the kinetic energy of a high-velocity fluid and turn it into the rotational energy of a shaft. Impulse turbines operate at atmospheric pressure because the runner is not submerged in water. The water pressure drops as soon as the water ejected from the nozzle hits the turbine blades, transferring its kinetic energy to the turbine runner.

Pelton and Turgo runners generally represent large impulse turbines and can be used in both horizontal- and vertical-axis arrangements. However, the Turgo runner (Fig. 4.1) is difficult to manufacture on a large scale; thus, the Pelton runner dominates high-capacity installations. The large-scale Pelton runner can be arranged in a horizontal or vertical axis configuration (Fig. 4.2). The horizontal axis arrangement is limited by the number of water jet nozzles per runner, which are usually two; hence, the water discharge is limited, and so is the torque. Therefore, large-scale horizontal Pelton turbines are arranged with a dual-runner configuration when higher installed capacities are required. Whereas in a vertical-axis arrangement, the number of nozzles can range from four to six, thus the installed capacity could be higher. The vertical-axis Pelton runners are also ideal for modern partially subsurface powerhouses or RC caverns, making them suitable for mountain power plants that are often constrained by spatial limitations.

Figure 4.1 Examples of a pair of single-nozzle vertical Turgo units installed in Taiban mini hydropower plant (Southern Taiwan).
Štefan Tkáč (2024).

Figure 4.2 Examples of two Taiwanese hydropower plants with Pelton turbines, namely, Daguan hydropower plant no. 1, unit no. 5 using German Voith-made horizontal axis dual jet, dual Pelton turbine (*left*) and Eastern Powerplants—Bihai unit using Indian BHEL-made vertical axis, six jets Pelton turbine (*right*).
Štefan Tkáč (2013).

The recent advances in Pelton turbines have led to improved performance through the overall computational modeling of the flow, droplets leaving the buckets ("splash losses"), casing design development, introduction of multinozzle designs, bucket shape modeling using Computational Fluid Dynamics (CFD), and new material technologies. The goal of all these improvements is to maximize turbine performance by increasing efficiency, power density, and turbine application, and by extending the life span of the runner itself, particularly in cases where the Pelton turbine is used to retrofit existing hydroelectric stations.

Key technological advances

Advanced simulation and AI optimization. The applications of Computational Fluid Dynamics (CFD) and the Finite Element Method (FEM) are standard practices for modeling water flow dynamics, pressure distributions, and structural integrity. In the case of an impulse turbine, the flow modeling is essential as the goal is to test the interaction between the nozzle (water jet) and the bucket to understand how to improve the moment of the energy transfer ("impulse transfer").

Optimization algorithms: The hydropower engineers are utilizing artificial intelligence (AI), specifically multiobjective genetic algorithms (e.g., the Non-Dominated Sorting Genetic Algorithm (NSGA-II), which is widely used in hydropower for optimization or multiobjective problem solving, such as hydropower cascade planning (also multiple reservoir solution for varying flow parameters and operational hill chart expansion), to optimize bucket geometry (angle, length, width) for the greatest efficiency and minimum energy losses automatically.

Advanced simulation techniques: The Smoothed Particle Hydrodynamics (SPH) technique provides a more accurate representation of complex flow events such as jet/bucket impact and splash, which were previously difficult to recreate accurately with traditional methods.

Predictive analytics: With these digital techniques, an abundant number of virtual tests can be performed, and the standard design iteration time is drastically decreased, reducing the need for expensive physical prototypes and enabling shorter lead times in product development.

Innovative turbine designs

High-density horizontal turbines: Although the multinozzle system is specific for vertical arrangement, the first horizontal six-nozzle Pelton turbine has successfully completed trials (2022-2023 Gerlos 1 pumped storage power plant in Zillertal, Austria, manufactured by Voith Hydro) and represents a significant advance in turbine design. This configuration increases the power density and rotational speed of turbines and generators significantly and creates an opportunity to downsize and downweight turbine design for retrofitting existing power plants. This is beneficial because it means less extensive civil works will be required to replace aging power plants.

Multinozzle optimization: Studies have demonstrated that adding nozzles (i.e., increasing the number of nozzles from 1 to 3 up to 6) leads to greater efficiency by providing more consistent and uniform load distribution across the runner.

Condition monitoring and maintenance

Smart monitoring: Smart online and offline condition monitoring technologies use vibration signals and artificial neural networks (ANN) to identify potential damage to the turbine so that performance problems can be rectified before they become critical. These techniques are being developed to provide a real-time assessment of condition and performance issues that may occur while the turbine is in operation.

Predictive maintenance: Predictive maintenance is an emerging technique for monitoring equipment condition and troubleshooting issues. Using these techniques enables equipment operators to obtain complex data to determine when to service the turbine to extend the useful life of the turbine, even under challenging operating conditions.

Material science and manufacturing

New advanced materials: Research is being conducted on the development of new materials, including new martensitic stainless steels, to provide enhanced strength and resistance to the extreme stress/time cycles generated by hydrodynamic back-pressure and abrasion caused by sediment in the water flow.

Modern manufacturing: New manufacturing techniques like rapid prototyping and 3D printing are being explored to produce complex, optimized bucket shapes that were previously difficult to manufacture using traditional methods. In general, the Pelton turbine is the most complex to manufacture. Traditionally, it is not manufactured as a whole, except for some pico cases. Hence, traditionally, the runner's main body and buckets are manufactured separately and later assembled and balanced. A similar process is followed for Cross Flow runners. The only Impulse turbine runner that was ever cast or forged as a one-piece runner was the Turgo type. But as forging and CNC precision advance, many manufacturers are shifting toward producing the Pelton turbine as a one-piece, allowing for a tougher, more complex design. This technology was initially available to small-scale hydropower plants, but it is now being expanded to larger scales.

Flow control and casing optimization

Aeration and casing design: The design of the Pelton turbine (not exclusively—also Cross-Flow turbines) casing is being optimized using fluid dynamic studies to improve aeration, minimize ventilation losses, and better manage splash water, potentially improving efficiency by several percentage points.

Counter-flow technology: Although predominantly developed for reaction (Kaplan and Francis) rather than impulse turbines (Pelton), as these are not directly affected by tailwater fluctuation and work in the atmospheric pressure, in some specific cases, the counter-flow technology could actually affect the Pelton turbine. The projects, such as AFC4Hydro, where water is injected into the tailwater section (draft tube or tail pipe) to counteract swirling flows and pressure fluctuations, thereby increasing the service life, stretch the limits and thus expand the flexibility and, in theory, reduce the water consumption of the existing turbines under adverse loads as a reaction to more fluctuating flow, especially during the off-peak efficiency. With regard to the Pelton runners, more laminar flow within the tailrace allows for faster and steadier outflow. However, in the direct context of Pelton turbines, Counter-Flow Technology refers to the specific implementation of a so-called counter-pressure system, also known as the Counter Pressure Pelton Turbine (CPPT), which is specifically developed for closed systems or pressurized environment thus ideal for pipeline applications, for example, drinking water pipes or water treatment plants. The CPPT contains elements such as pressurized air cushion, water-air separation, or active regulation via PLCs to balance the water level and air pressure in the casing, preventing the turbine runner "drowning."

Reaction turbines

Reaction turbines, which refer to hydropower runners, use fixed or adjustable blades specially designed to harvest both kinetic and pressure energy from the water flowing through the runner itself as the turbine is working analogously to pump the water, it is operating in reverse, thus it is called a reaction as it literary reflects Newton´s third law of motion *(for every action (force), there is an equal and opposite reaction).*

Regarding energy conversion, reaction turbines capture the kinetic energy of a low- or high-velocity fluid, usually flowing into the turbine via a spiral casing, thereby creating vortex-like conditions before entering the runner. In contrast to Impulse turbines, the reaction turbine runners operate fully submerged in the water environment, and their tailrace-draft tubes must ensure that no air can enter the system, as negative pressure caused by suction occurs behind the turbine blades. The water pressure changes from positive to negative as it passes through the turbine, transferring its kinetic energy to the turbine runner. Suppose the water pressure drops too fast after the runner. In that case, reaction turbines might develop adverse effects called cavitation, causing erosion/pitting, harsh vibration accompanied by noise, and eventually leading to turbine performance losses.

Francis

The Francis Turbine is one of the first modern, large-scale water turbines and was designed by British-American engineer James Bicheno Francis in 1949. Since then, this turbine has undergone many design modifications and is now regarded as the most adaptable and effective hydro turbine runner, making it the dominant design of turbine runners in operation throughout the world. Approximately 60% of the total hydropower capacity in the world is produced by Francis Turbines. A recent installation by both Dongfang Electric and Harbin Electric of 16 × 1000 MW Francis Turbines at Baihetan Hydropower Station, represents the highest capacity for a single Francis unit anywhere in the world.

The demand for renewable energy technology in all capacities has encouraged a shift to Francis Turbine design, to create more adaptability to variable renewable energy resources, including wind and solar, reduce flow instabilities, maximize energy efficiency across a wider range of operation, and create increased durability using advanced modeling techniques and new materials. To support this movement, the runner design of Francis turbines is tailored to the head and flow conditions specific to a given site based on the specific speed of that turbine (N_s). There are three specific speed categories for Francis turbine runners: low specific speed with radial flow, high heads, and low flows ($N_s<50$); the normal/medium or average specific speed, where runners usually demonstrate a mixed flow configuration, moderate heads/flows (N_s approximately 50 to 90)); and high specific speed runner designs demonstrate axial flow and low heads/high flows ($N_s>90$). Looking at the evolution of the Francis turbine runner designs from low specific speed to high specific speed shows an evolution of the design philosophy behind the construction of hydropower turbines (Fig. 4.3). When designing the runners for Low specific speed Francis turbines, designers use predominantly radial configurations that work best for high head environments. As the specific speed of the turbines increases, so does the axial portion of the turbine runners, thus, making it possible for the design to process significantly larger flows within the system. The evolution of design demonstrates the improvement in energy capture from the turbines, which can be realized by working with different hydrological conditions experienced in the rich energy carried by water flow.

Enhanced operational flexibility: The main focus of ongoing research is increasing the efficiency and stability of fully developed Francis turbines when operated across a much larger operating range, from a very low turbined volume to maximum. This means reducing the excessive vibrations and excessive pressure fluctuations that occur at a Francis turbine's operating levels below and above the Best Efficiency Point (BEP), which have always limited the existing operating envelope.

Figure 4.3 Turbine open-air museum at the Guishan Hydropower Plant, Taiwan, shows examples of the three standard Francis runner designs tailored to the runner's flow, head, and specific speed (N_S). (A) Fast (low head), (B) medium (medium head), (C) slow (high head).
Štefan Tkáč (2025).

Computational fluid dynamic (CFD) and fluid structure interactions (FSI) simulation technology: Researchers use advanced CFD and FSI modeling and simulations to accurately understand: (1) flow patterns through a Francis turbine and the resultant pressures acting on the runner and draft tubes; (2) the internal flow dynamics within a Francis turbine; and (3) the forces being exerted on and through the various components of the turbine and, thus, be able to create and verify new geometries for the runner and draft tube without excessive use of prototype testing.

Instability mitigation techniques: New technology is being created to counteract flow instability through innovative hardware and operating approaches.

Vortex rope suppression: Air/water injection into draft tubes or using circular rods/antiswirl fins to separate and weaken larger, more dynamically unstable vortex systems, which reduces severe pressure pulsations and power swings.

Runner blade optimization: The configuration of blade profiles and the use of splitter blades (shorter blades located between full-length blades) can minimize fluid dynamic instability, thereby providing an improvement in cavitation performance and reducing rotor-stator interactions (RSI).

Material science and erosion resistance: Both the study of material science and erosion resistance is being investigated to develop new materials and protective coatings (including "hard coatings" and specialized weld surfacing alloys) that can provide improved resistance to cavitation and sediment abrasion (the two primary sources of problems related to sedimentation) in rivers with high in situ silt concentrations (e.g., the Himalayan river system).

Variable speed technology for pumped storage hydropower plant applications: The technology of variable speed generators, including doubly-fed induction generators and full-sized frequency converters, allows pumped storage hydropower plants to more precisely control the speed of turbine rotation, ultimately improving the efficiency of operation of these plants by approximately 3% below and at low heads and partial loads while broadening the operational range of the equipment.

Advanced monitoring and diagnostics: Intelligent condition monitoring systems utilizing pressure, acceleration, and acoustic sensors, as well as novel fiber-optic sensing, are being developed and utilized to provide real-time fault diagnosis and predict when maintenance will be necessary on various components of pumped storage hydropower plants to prevent failures and prolong their operational lifespan.

Fish-friendly designs: "Fish-friendly" design is a rapidly growing area of research on improving the survival of fish when passing through a Francis turbine by reducing the occurrence of large pressure changes and high acceleration events. Examples of the fish-friendly design technologies include "Sensor Fish" devices used for in situ characterization.

Kaplan

Recent advances in the design of Kaplan turbines have revolutionized the hydropower industry, creating a demand for more efficient, lower-impact, and more flexible hydropower systems. With renewable energy sources growing rapidly in popularity around the world, these developments will allow for improved energy solutions that are both environmentally and economically sustainable.

Advanced simulation and optimization The Hydropower Engineers utilize advanced Computational Fluid Dynamics (CFD) and Finite Element Analysis (FEA) to develop and optimize blade geometry and flow paths in a virtual environment. This enables the rapid testing of a larger number of blade configurations, increasing the output and efficiency of turbines and allowing for a more accurate prediction of the stresses that will occur and potential failure points prior to the manufacture of physical prototypes.

Environmental compatibility

Oil-free hubs: represent a significant design move that companies have been developing since the 80 s. The main idea is to replace the traditional pressurized oil used as the hydraulic fluid to move the adjustable parts of the turbine, such as turbine blades or additional wicket gates, with biodegradable, eco-friendly lubricants, self-lubricating materials (Quaranta, 2023), or even water, thereby reducing the risk of water contamination in the event of leakage, allowing for easier maintenance and in some cases lowering the friction (Voith, 2026; Andritz, 2026).

Fish-friendly designs: Through active research and innovation, a turbine design that minimizes fish injury during turbine passage has been developed through the use of minimum gap runners (Voith, 2026; Andritz, 2026).

Energy storage through hybridization

Kaplan units will be combined with battery energy systems to provide additional battery power during times of FCR (frequency control reserve) to stabilize the electricity grid frequency. This will reduce the number of constant adjustments that turbine guide vanes and runner blades must make, thereby reducing the mechanical wear and tear associated with the regular operation of guide vanes and runner blades.

Increased operational flexibility

"Smart" Operations: Field/field testing has provided the basis for the development of "synergy curves." Data-driven techniques (and sometimes a combination of data-driven methods and machine learning) have enabled operators to operate their turbines in a safe and effective manner across an expanded operating range of load and water flow.

Counter flow technologies

New technologies, such as turbine diverter designs that inject additional water into the turbine, or the use of mechanical devices to provide counterflow and/or pressure relief downstream of the turbine, are being developed so that existing turbines will have an increased ability to operate safely and efficiently for longer periods of time.

Modular and compact designs

New designs (e.g., the Siphon TM, Kaplan EVO, and rim-drive generators) are being created to develop a simple, efficient, modular, compact energy generation system to eliminate the need for large power plants (e.g., civil structure) and dramatically reduce the amount of time, funds, and environmental impact involved in constructing new power plants. These systems will also be easily adaptable for use in older power plants (e.g., runoff).

Alden turbine

Alden turbine is a relatively new model that focuses on cutting-edge efficiency, retrofitting older Francis or Kaplan units, and has become famous for its fish-friendliness, with fish survivability rates up to 98%. It has been developed by the German company Voith with the support of the Electric Power Research Institute (EPRI) and the Department of Energy, all in direct cooperation with the Alden Research Laboratory, Inc., the oldest continuously operating hydraulic laboratory in the United States (Voith, 2026).

So far, the Alden runner has not been commercially applied, but it has been tested in a scaled form. The known configuration is based on a vertical shaft, and despite the Alden turbine being famous for slow speeds and a three-bladed helical design, there is no information on its application in reverse mode. Hence, the Francis and Kaplan turbines remain leaders in versatility and application range.

Pumped storage

The current development in pumped storage technology revolves around two major design concepts: Variable-speed Control technology and Ternary set technology.

Variable-speed control

As RES technology has many applications on the grid, there will be a need to stabilize the grid because of the large variation in the use of RES on the grid. In response to this challenge, variable-speed technology was created to allow for an effective way to stabilize the grid with a renewable energy mix saturated with variable sources. The variable-speed pumped storage facility provides the ability to adjust the rotational speed of the turbine in reversible operation using an asynchronous or synchronous (with a frequency converter) motor-generator. This allows turbines to be operated within a larger operating envelope and permits the pump's output to be adjusted according to the amount of energy available for storage at any time.

Ternary set technology

A ternary system includes three main pieces of equipment: a motor/generator, a separate turbine (generally either Francis or Pelton), and a pump assembly. Due to utilizing a common motor/generator to rotate both hydraulic systems in the same rotational direction during both operational states, the overall economic efficiency of the power generation facility can be maximized. There are multiple ways to transition from turbine to pump operation within a ternary system, including the use of a clutch that can be engaged while stationary, a starting turbine, or a torque converter that can synchronize both systems. The configuration also creates a hydraulic short circuit between the two hydraulic machines, which provides the best opportunity for a quick response to changing electrical grid conditions. The torque converter allows rapid switching between the pump and turbine modes of operation. Both operational modes create full regulation of the power output of the unit from 0% up to 100%.

Francis turbine

Francis turbine is a standard reversible turbine runner even in the case of ternary sets, where it can be applied as a multistage Francis turbine pump; however, as the ternary sets combine the pumping sets and turbine set separately via a torque converter and clutch, a vertical shaft Pelton runner may be applied in turbine mode.

Low-head pumped-storage systems, usually up to 200 m, can also use Kaplan or Deriaz turbines, or a combination of a Francis turbine and a Kaplan turbine, due to versatility. Kaplan and Deriaz turbines in reversible mode rely solely on the runners' ability to adjust their blades to regulate the runner's revolutions.

Small scale

Technological improvements in small hydropower turbine technology make it possible to now create decentralized hydropower systems with a greater range of applications, efficiencies, and adaptable features than was previously possible. Small-scale hydropower systems have significantly improved due to technological advancements, including new materials, higher precision manufacturing, better flow modeling, new smart digital control systems, and so forth, which allow for the operation of more efficient turbines in a wider variety of hydraulic conditions. Impulse turbines are primarily used at higher elevation sites with lower flows; the Pelton and Turgo turbine types have been extensively tested and refined since their inception to optimize bucket geometry, develop advanced injection designs, and utilize advanced wear-resistant alloys and coatings that improve the life of impulse turbines in very abrasive situations. On reaction turbine technology, both the Francis turbine and Kaplan turbine have also undergone many major improvements and innovations since they were first developed. These improvements include compact design, variable runner configurations, improved cavitation performance, etc. Therefore, both the Francis and Kaplan turbines are suitable options for medium to low elevation applications where small amounts of hydropower can be generated.

The newest entry to the list of turbines is the Siphon turbine, which allows run-of-river and low-head retrofits to be made without a large amount of civil construction. The fact that it can be operated on top of the dam using siphonic pressures gives it an opportunity to provide a low-impact and cost-effective way to support modern ecological and regulatory initiatives. Together, these technologies allow small hydropower to develop into a more versatile, environmentally aware, and economically sustainable part of the global renewable energy mix.

Fish friendliness

Recently, small-scale hydropower technologies that are friendly to fish have developed exponentially due to increased awareness of the environment, the importance of protecting biodiversity, and the regulations for environmental protection that have become prominent during the planning for hydropower development. Turbines of today increasingly seek to maximize both energy generation and safety for migratory fish, decrease fish death rates, and limit their effects on habitats around the turbine site. The development of Vortex turbines has become one of the greatest achievements in turbine technology, as these turbines create energy using controlled vortex flow patterns, allowing for low-speed rotation and smooth blade shapes. The hydraulic environment in which fish pass through these turbines is gentle, reducing the likelihood of injury to fish and making them ideal for use in small rivers that are home to sensitive ecological systems.

Another type of fish-friendly turbine is the Archimedes Screw Turbine (AST). The low rotational speed of ASTs and the large distance between blades allow fish to swim freely through the screw without being harmed; thus, ASTs are very efficient energy-producing devices, particularly in areas with little elevation difference between high/low water (i.e., head loss). DIVE turbines are another option for producing energy via hydrokinetic forces, but they are more compact and vertically oriented than ASTs. DIVE turbines were designed specifically to reduce the chances of fish being struck by blades or subjected to sudden shifts in water pressure while swimming through a dam or hydroelectric plant; therefore, DIVE turbines will not be as compatible with species of fish that migrate up rivers/streams.

With regards to future advancements, Alden turbine technology appears promising for future small hydropower applications. The Alden turbine was designed initially for very large-scale applications using three blades to maximize efficiency while minimizing risks to fish. By optimizing the pressure levels through carefully controlled pressure regimes, the Alden turbine will demonstrate that energy efficiency and sustainable ecological performance are compatible.

If the Alden turbine can be successfully scaled to small hydro applications, it will represent a new generation of biologically optimized small hydro turbines developed with a low environmental impact.

The ongoing research, monitoring, and field testing of fish-friendly turbines will increasingly occupy an important role in sustainable hydropower development around the world.

Retrofitting

Modernization of existing hydropower facilities is now at the heart of development for small hydropower facilities. Therefore, modernization efforts are underway to improve the efficiency and lifespan of existing hydropower stations, as well as to meet stricter environmental regulations, without the need for costly capital investment in major civil reconstruction. As Turbine technology advances, hydropower engineers continue to incorporate new and/or improved turbine designs into existing facilities; therefore, these upgrades ultimately provide the required performance benefits while preserving a facility's existing infrastructure.

A prime example of this trend is the dual-runner Francis Turbine Technology. Dual-runner turbines utilize twin runners located within a single Turbine housing. This unique application provides improved efficiency over an expanded flow range, making it an exceptional and flexible choice for retrofit applications in existing plants where flow variations have increased due to climate change and/or changing water management practices.

The Reintroduction of the Camelback Turbine, historically one of the "workhorses" of Low-Head Hydropower, is another prime example of the basic turbine designs being revised using advanced engineered materials, precision casting technology, and optimized hydraulic profiles. Newly engineered Camelback Turbines can deliver superior reliability and performance to existing Heritage Plants where structural constraints have prevented a major turbine replacement.

Retrofit technologies are also being developed that utilize increasingly specialized forms. For example, the Kennelbach-Schindler turbine was originally designed for operation in very low head situations, but many hydraulic designs have now been made, and the turbine's compact modular design and increased efficiency make it an attractive option for refurbishing and upgrading very old low head stations. Another example of a turbine that offers similar capabilities is the DIVE turbine, which was designed with an environmentally-conscious submerged design to ensure that fish and other aquatic life are minimally affected by its installation in the water. The DIVE turbine can generate electricity in shallow-water locations, opening up many opportunities to use previously unexploited areas of older canals and smaller dam sites to generate power.

While both the Kennelbach-Schindler (Escher-Wyss-made Quadruple-Francis turbine) and DIVE turbines focus on low-head applications, the Siphon turbine is applicable to vessels being retrofitted for use without extensive civil work. A Siphon turbine can be placed above the dam crest so that the flow is siphoned through the turbine. By taking this approach to turbine placement, it is possible to retrofit run-of-river dams or facilities without any major alterations to the existing infrastructure, allowing the plants to increase their generation capacity or to restore their original operating efficiencies.

Collectively, these new technologies are introducing new possibilities for the rehabilitation of small hydropower plants, providing new avenues for engineering with the use of old infrastructure, as well as preserving small hydropower industrial heritage and contributing to the field of renewable energy with the many new options available.

Kinetic turbines

Hydrokinetic technology harnesses the kinetic energy of flowing water directly, without the need for a dam. Devices such as underwater turbines and oscillators convert the kinetic energy of flowing water, or even a current, into electricity. Therefore, the technology, although predominantly designed for the ocean currents and tidal power plants, is also applicable to the river streams. To better adhere to stream properties, in some cases, biomimicry following the shape and movements of shark fins is adopted (bioSTREAM) (Knight et al., 2014).

List of abbreviations

ALDEN	Alden Research Laboratory, Inc. (Alden)—the oldest hydraulic laboratory in the United States
AST	Archimedes Screw Turbine bioSTREAM—Biomimetic Stream Energy Device designed and developed by an Australian company, BioPower Systems Pty Ltd.**BEP** Best Efficiency Point
CPPT	Counter Pressure Pelton Turbine CFD—Computational Fluid Dynamics
Deriaz turbine	a diagonal-flow reversible turbine invented by Paul Deriaz in 50 s DIVE turbine - designed by the German DIVE Turbinen GmbH & Co. KG as a fish-friendly, compact, and vertically oriented hydrokinetic turbine.
FEA	Finite Element Analysis FCR—Frequency Control Reserve Francis Cammellback - Dual Francis turbine mounted on a single shaft, hence a dual action Francis (For more power, the systems could be doubled).
FSI	Fluid Structure Interactions IEA—International Energy Agency
Kennelbach-Schindler	Quadruple-Francis turbine (Quadruple action system similar to Dual action Francis Cammellback design, but multiplied)
NSGA-II	Non-Dominated Sorting Genetic Algorithm II PLC—Programmable Logic Controller RES - Renewable Energy Sources RSIRotor-Stator Interactions SPH—Smoothed Particle Hydrodynamics

References

Andritz. (2026). Fish-friendly turbine technology [PowerPoint slides]. Andritz Hydro, Retrieved January 13, 2026, https://www.andritz.com/resource/blob/242614/5c5c514cf7fe4947bcac231ce81d98f6/10-presentation-fish-friendly-turbine-data.pdf.

Knight, C., McGarry, S., Hayward, J., Osman, P., & Behrens, S. (2014). A review of ocean energy converters, with an Australian focus. *AIMS Energy, 2*(3), 295–320. https://doi.org/10.3934/energy.2014.3.295.

Quaranta, E. (2023). Lubricant oil consumption and opportunities for oil-free turbines in the hydropower sector: A European assessment. *Energies, 16*(2), 834. https://doi.org/10.3390/en16020834.

Voith. (2026). Environmentally friendly turbine design. [Brochure]. Retrieved January 13, 2026, from https://www.voith.com/corp-de/Eco-friendly-turbine-design.pdf.

5

Architecture and design

Introduction

From the architectural point of view, the hydropower plants are very special objects that are designed to be functional and protective on one hand, but on the other hand, typologically they probably represent the most intense sense of nature, literally flowing through them. The fact that there are not two identical hydropower plants built raises the technical and historical value of individual cases, making them a perfect subject of industrial heritage studies as well as a unique site-associated reference for future projects.

Also, the architecture of hydropower plants represents the most durable man-made Civil engineering structures. This is proven by numerous industrial heritage cases around the globe. An interesting example of this could be the remains of the annually flooded Sogi Power Station from 1909 (Mainichi, 2021) located in the city of Isa, Kagoshima Prefecture, which is always fully visible once the Tsuruta Dam water level drops to approximately 120 m (MLIT, 2026).

The aspects of durability of the hydropower plants, along with the need for human interaction to operate them, were also popularized in the popular visionary/dramatic television series "*Life after People*" by David de Vries (IMDb, 2008), where hydropower plants were depicted as the most durable and independent civil engineering structures, which would most likely be the last to perish, should people suddenly disappear.

Although the development of the interior of the hydropower architecture has been tilted toward utilitarian and minimalistic styles, making more room for the modern hydropower technology, which in contradiction requires still less and less space, the façade, material solution, and the overall shape are still spheres where architects kept their freedom for interaction with the surroundings. An excellent examples are the Hydroelectric Power Station by Becker Architekten, built on the river Iller, in contrast to the old industrial architecture in Kempten, Germany (Architizer, 2018), and more ecologically shaped Hydroelectric Power Station in Punibach Bolzano, Trentino-Alto Adige/South Tyrol, Italy designed by Monovolume architecture + design, designed to respect the natural realm

Hydropower Plants. DOI: https://doi.org/10.1016/B978-0-12-824520-0.00007-1

and merge with the mountainous landscape of the region (Architizer, 2013).

General parts of the hydropower plant

Despite being depicted as mature technology and often referred to as simple industrial buildings equipped with primitive energy-conversion machinery dating back to the 18th century and beyond, once studied deeper, one can quickly realize that the hydropower plants are in fact complex systems composed of a sequence of carefully coordinated structures that guide water from the river to the turbine and safely return it downstream. At the upstream end, diversion structures such as weirs or dams and intake facilities regulate flow and protect the plant from debris and excessive sediment. Water is then usually conveyed through the headrace, which may take various forms, such as an open canal, a tunnel, or a combination of both, often incorporating desilting basins to remove silt, spillways to adjust the flow are usually part of the basin as well as the next forebay tank, which all together with desilting basin ensures hydraulic stability and water quality. In middle- and high-head systems, pressure fluctuations may occur due to rapid load changes; thus, a surge tank is introduced as a critical safety and control element. From there, the penstock delivers water under pressure to the turbine house, serving as the system's primary energy-transfer link. The turbine house integrates mechanical and electrical equipment, including turbines, generators, exciter, governor (HPU—hydraulic pressure unit), flywheel, shaft, couplings, and emergency brake to form a complete generation set (known as a *gent-set*). Supporting infrastructure such as cranes, maintenance systems, control rooms, and switchyards enables reliable operation and grid connection. After energy extraction, water exits through the draft tube, recovering pressure and improving efficiency. Finally, the afterbay and tailrace canal return the water to the river, completing the hydropower cycle.

To illustrate the entire hydropower cycle, a further generalized list of hydropower plant elements is provided, arranged hierarchically from the inflow to the tailrace. The hierarchy of each element might vary slightly depending on turbine type, head ratio, and site preferences.

- **Diversion structures** *(diverting water to headrace—could be **naturally formed** or **designed**)*
  - **Weir or dam** (adjustable or fixed—usually with crest overflow)
  - **Intake structure** (protected by two sets of debris elimination:

  - **Coarse trash racks** (bar racks—designed to capture large debris)
  - **Fine trash racks**—screen (installed right in front of the turbine intake to catch smaller particles—often equipped with an automatic cleaning system—they are installed in the forebay tank)
- **Headrace** *(diversion canal, tunnel, or combination of both)*
  - **Settling/Desilting basin** *(optional but necessary in case of sediment-laden flow)*
    - Spillway
  - **Forebay tank**
    - Spillway
    - Intake opening—protected by Fine Trash Racks
  - **Surge tank** *(installed in case of middle and high head designs)*
  - **Penstock** *(longer and visible in case of middle and high head designs, short and integrated with the powerhouse in case of low and some middle head designs)*
- **Turbine house** consists of **turbine hall** or Turbine/generator **installation platforms** in case of vertical arrangements, and **control room** and **additional spaces** for operation *(shape is based on size, gen-set type, and preference of the surroundings).*
  - **Maintenance systems**
    - Crane *(usually overhead type—electric or manual hoist, in case of SHPP with a hoist capacity up to 10 tons). An external mobile crane is also possible, but an appropriate roof opening must be designed.*
  - **Control room** *(the size and location depend on the size of the powerhouse—some systems are integrated within the turbine hall)*
  - **Switchyard** *(outside, but within the premises of the powerhouse)*
  - **Gen-set** *(vertical or horizontal axis arrangement)*
    - Exciter
    - Generator
    - Flywheel *(usual part of the impulse and vertical axis units, position in the gen-set depends on design, but the flywheel is either placed between the generator and turbine or at the end of the turbine, accompanied by the emergency break)*
    - Shaft with couplings
    - Turbine
    - Governor (HPU)
- **Tailrace**
- **Draft tube or tailrace tunnel** *(in case of impulse turbine, not necessary. A sluice gate protects the tailrace outflow opening, and sometimes also coarse trash racks are applied against debris from typhoons and floods, which may cause blockage).*

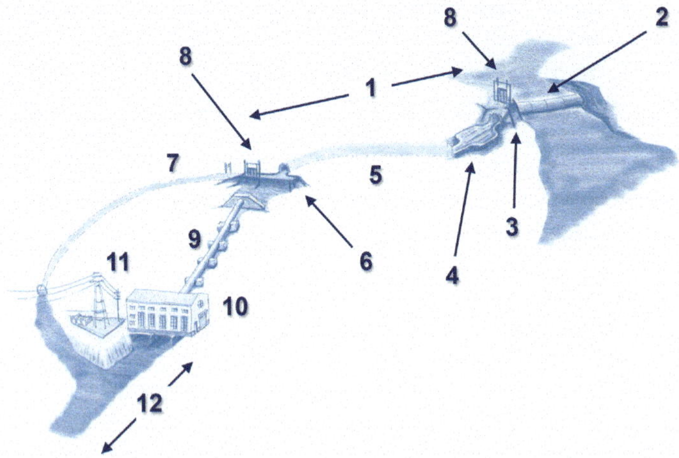

Figure 5.1 Parts of the hydropower plant (headrace (1), diversion structure (inflatable weir) (2), intake structure (3), settling/desilting basin (4), diversion tunnel (5), forebay tank (6), spillway (7), intake gates (8), penstock (9), turbine house (10), switchyard (11), tailrace (12)).
Author's personal drawing.

- **Afterbay** *(depression is usually designed for the reaction turbines to maintain the pressure)* (Fig. 5.1).

Typology, hierarchy, and architectural styles

The very first hydropower technology was fully mechanical, and thus most of the governance and electrical wiring was fully exposed. The reason for open technology was that most of the mechanics were transferred from the improved horizontal-axis waterwheel, and that it was much easier to maintain the entire gen-set on a single level rather than to move between levels. Another reason for the open technology was cooling for the large-scale and often overheating generators. In fact, the first power plants were prone to fires, and there are many cases that support this statement. Additionally, the architecture was mostly serving as a necessary cover and thus was made fully from wood or in combination with stone-based masonry.

The ventilation was therefore the main factor that led to the development of certain features that could identify the typology of the hydropower plant Architecture even today.

Since the early generators did not have their own cooling nor fan, the only way to handle the temperature was to create an artificial draft inside the building. This was achieved by the installation of narrow but long vertical windows usually with slots or louvers on the bottom and top. These were initially fixed and later adjustable. The inner space was also using the stack (chimney) effect by installing a monitor top which would harvest the energy of the passing air and draw the hot air from under the top of the roof. Another solution was to simply draw in the cool air from the bottom of the building where the tailrace was usually located, and the air was cooled by the flowing water. In some cases, the wooden floor was acting as the only membrane between the tailrace and the machinery room, hence the air naturally flows through the gaps between the floor construction.

As hydropower technology got vertical, the need for adaptable architecture brought a new typology for the surface constructed power houses. The new partially subsurface power houses emerged. These new types of structures strictly divide the water part from the electrical within the gen-set. The inflow penstock, the main valve, casing, water turbine and the draft tube were all underneath while the generator and exciter were above the surface for easy access and maintenance. This vertical arrangement could protect the power plant to a certain level in case of flooding and allows the project to fully capitalize the potential of the head.

With the use of most of the ideal location for the surface and partially subsurface hydropower houses, the engineers start to explore more challenging solutions that would allow to build a hydropower plant resistant to weather and to a certain extend also seismic activity. The solution is called a cavern type powerhouse, and it is generally projected in a vertical configuration with a pair of separated caverns, one for gen-set/s and one for the switchyard. The entire facility is placed underground hence shielded from any sudden weather conditions, landslides, typhoons, or potential danger of attack *(sabotage of the energy infrastructure of regional or even national importance)*. The cavern adopts a vault or domelike structure, which is common structural elements showing good resistance to earthquake tremors. The cavern is also fully embedded in the mountain; therefore, it adopts the same frequency during the seismic activity thus it's less prone to collapse in contrast to more vulnerable surface or partially subsurface types (Fig. 5.2).

Industrial architecture and its connection to hydropower plants

Industrial architecture emerged as a direct response to the needs of production, infrastructure, and technological advancement. Hydropower plants represent one of its most distinctive and enduring expressions, where engineering requirements, environmental constraints, and architectural form are inseparably linked. Unlike other building typologies, hydropower facilities are shaped primarily by the movement of water, the scale of machinery, and long-term operational resilience, resulting in architecture that is both highly functional and symbolically powerful.

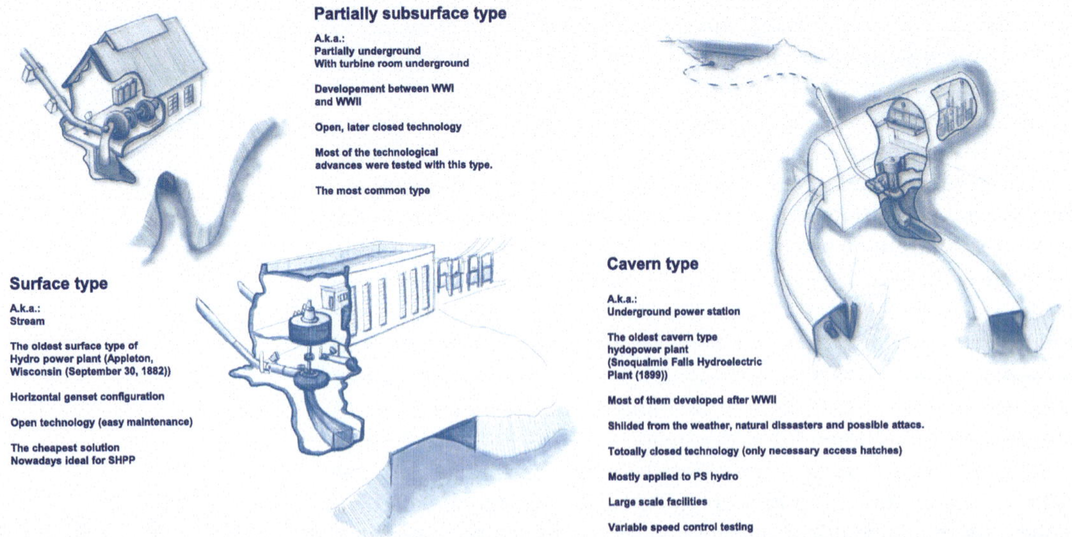

Figure 5.2 Evolution of the power houses.
Author's personal drawing.

Core characteristics of industrial architecture in hydropower plants

A defining principle of industrial architecture is the **integration of functionality and design**. Hydropower plants epitomize this approach: their spatial organization, structural systems, and material choices are governed by operational efficiency, yet they often achieve a monumental and expressive presence. Exposed structural elements, visible machinery, and clearly legible construction logic convey strength, permanence, and technological progress rather than decorative refinement.

Material selection plays a central role in this architectural language. **Raw and robust materials such as reinforced concrete, brick, and steel** dominate hydropower construction. Concrete, in particular, offers exceptional compressive strength, water resistance, and durability under dynamic loading, making it indispensable for dams, spillways, and powerhouse structures. Brick and metal elements contribute to structural integrity while reinforcing the honest, utilitarian esthetic typical of industrial architecture.

Internally, hydropower plants are characterized by **open floor plans with wide structural spans**. Turbine halls are designed to accommodate massive generators, turbines, and control equipment, requiring unobstructed spaces and reinforced walls capable of supporting overhead crane rails. These cranes impose significant dynamic loads during maintenance operations, influencing both structural design and spatial proportions. The openness of these interiors also facilitates ventilation, inspection, and the movement of heavy machinery throughout the plant's operational life.

Closely related to this openness are the **high ceilings** typical of hydropower facilities. Vertical clearances are dictated by the size of turbines, generators, penstocks, and lifting equipment. These soaring interior spaces reinforce the monumental character of the buildings while serving a strictly utilitarian purpose.

Externally, hydropower plants often exhibit a **lack of ornamentation**. Façades are generally plain, with form following function and structural clarity prioritized over decorative elements. Over decades of operation, exposure to water, vibration, and weather leads to **distressed and worn finishes**, which further enhance the authentic industrial character of these structures. Rather than being concealed, signs of aging often become an integral part of the building's identity.

Another hallmark is the **exposure of brickwork, ductwork, piping, and mechanical systems**. Large-diameter pipes, steel frameworks, and electrical installations remain visible, especially in older plants, where accessibility and ease of monitoring were essential. This "open technology" approach allowed operators to quickly identify faults and perform repairs, reinforcing the transparency and honesty of industrial design.

Windows also play a functional and esthetic role. **Large metal-grid windows** are common in historic hydropower plants, admitting daylight into expansive turbine halls and reducing reliance on artificial lighting. They support natural ventilation while contributing to the visual rhythm of industrial façades.

Together, these features define hydropower plants as archetypal examples of industrial architecture—structures where technical necessity and architectural expression are inseparable.

Flooding, embankments, and architectural adaptation

Hydropower plants are intrinsically linked to fluctuating water levels and flood events. Over time, repeated flooding has influenced the **evolution of embankments**, leading to higher, reinforced, or reprofiled protective structures around powerhouses and auxiliary buildings. These adaptations reflect accumulated operational experience and changing hydrological conditions.

Flood levels are often documented directly on the architecture itself. **Marks on walls, columns, or exterior surfaces** record historical flood heights, serving as both technical references and tangible reminders of extreme events. These markings form an informal but powerful layer of architectural memory, embedding hydrological history into the fabric of the building.

Architectural styles of the Industrial Age and beyond

While functionality dominates hydropower design, architectural styles of different periods have left distinct imprints on power plant esthetics.

Baroque influences, though rare, appear in early hydropower facilities designed to project national pride or industrial prestige. These structures emphasize symmetry, monumentality, and occasionally decorative façades or ceremonial entrance spaces, translating Baroque grandeur into an industrial context.

The rise of hydropower in the late 19th century coincided with **Victorian industrial architecture**. Powerhouses from this era feature robust brick construction, iron and steel detailing, and large windows. Ornamentation, while present, remains subordinate to structure and machinery, exemplified by early facilities such as those at Niagara Falls.

The advent of **Modernism** marked a decisive shift toward functional clarity. Influenced by Bauhaus principles and architects such as Le Corbusier, hydropower plants embraced reinforced concrete, clean lines, and minimal decoration. Structures like the Hoover Dam powerhouses exemplify the Modernist credo that form follows function, celebrating materials rather than concealing them.

The **Bauhaus movement** further reinforced rationalism and standardization. Bauhaus in hydropower was influenced by Jugendstil (Art Nouveau), proto-functionalism, and even Neoclassicism. A good specific example can be the Hammeren Hydroelectric Power Station. Bauhaus style hydropower architecutre further adopts a display flat roofs, modular forms, and unadorned façades, aligning industrial production with architectural discipline.

In contrast, **Neoclassicism** reintroduced historical references. Early 20th-century hydropower projects, particularly civic infrastructure initiatives, employed columns, symmetrical layouts, and monumental façades to convey authority and permanence, as seen in facilities associated with the Tennessee Valley Authority and the Carlo Esterle plants.

Regionalism sought harmony between hydropower architecture and local landscapes. Scandinavian power plants, for example, integrate stone, timber, and subdued forms that reflect regional materials, climate, and cultural identity.

Art Nouveau and Art Deco introduced ornamentation rooted in modernity. The Štvanice power station in Prague reflects Art Nouveau's fluid esthetics, while Art Deco later emphasized geometric forms, symmetry, and stylized decoration. The Hoover Dam's sculptural details exemplify this fusion of technology and artistic expression.

In colonial contexts, hydropower plants often adopted **colonial architectural styles**, blending European neoclassical elements with local materials and climatic adaptations, symbolizing technological progress alongside imperial authority.

Post-war reconstruction introduced new directions. **Modernism and postmodernism** allowed experimentation, symbolism, and contextual responses, as seen in facilities such as Ptusza and Hammerfest, where reconstruction architecture merged tradition with functional modern design.

Contemporary hydropower architecture increasingly reflects **Hi-Tech principles**, exposing advanced engineering systems and celebrating automation, transparency, and performance. **Modular architecture** enables prefabrication and scalability, particularly for small and remote hydropower installations. Looking forward, **parametric design** and computational tools allow hydropower structures to respond precisely to hydrodynamic and environmental conditions, optimizing performance while achieving sculptural integration with the landscape.

Finally, **ecoarchitecture and sustainability** define the present and future of hydropower design. Emphasizing ecological integration, circular economy principles, fish-friendly turbines, reduced civil works, and locally sourced materials, modern hydropower plants aim to balance energy production with environmental stewardship.

Conclusion

The architectural style and design of hydropower plants have changed substantially since they first appeared, in line with the various stages of the industrial revolution and the architectural styles that have influenced all aspects of our lives today. Hydropower plants can be said to reflect the evolution of industry and technology; the many forms of architecture that have accompanied these advances, from the splendor of Baroque to the economical, simplistic form of Modernism and Minimalism to the flexible form and design characteristics of Regionalism and Parametricism, all demonstrate how hydropower facilities incorporate function, form, and style, while using the power of nature for energy conversion. As such, hydropower facilities are a fusion of the architecture-environment relationship, in which hydraulic civil structures have been constructed as the only truly dynamic buildings, where "the nature literally flows through" throughout their entire designed lifespan.

List of abbreviations

HPU	Hydraulic Pressure Unit
MLIT	Ministry of Land, Infrastructure, Transport and Tourism
SHPP	Small Hydropower Plant
TVA	Tennessee Valley Authority

References

Architizer. (2013). *Hydroelectric Power Station Punibach Bolzano, Trentino-Alto Adige/South Tyrol, Italy.* Retrieved January 6, 2026, from https://architizer.com/projects/hydroelectric-power-station-punibach/.

Architizer. (2018). *Hydroelectric Power Station, Kempten, Germany.* Retrieved January 6, 2026, from https://www.qsr.mlit.go.jp/turuta/go/sogi.html.

IMDb. (2008) *Life After People (TV Movie 2008) ★ 7.3 | Documentary, Sci-Fi.* Retrieved January 6, 2026, from https://www.imdb.com/title/tt1173907/.

Mainichi. (2021). *Vestige of a hydro plant in a southwest Japan dam lake partially destroyed by downpours.* Retrieved January 6, 2026, from https://mainichi.jp/english/articles/20210713/p2a/00m/0na/024000c.

MLIT. (2026). *Sogi Power Station Remains - Tsuruta Dam Management Office | Ministry of Land, Infrastructure, Transport and Tourism, Kyushu Regional Development Bureau.* Retrieved January 6, 2026, from https://www.qsr.mlit.go.jp/turuta/go/sogi.html.

6

Global development trends

Hydropower development on a global scale continues to grow, as countries look for reliable, low-carbon energy sources. Large hydropower continues to be a key means of generating electricity in regions of the world, most significantly in developing nations where the need for more electricity is leading to the creation of large-scale projects. Small hydropower is becoming increasingly popular around the world as an appropriate means of electrifying rural areas and providing a renewable source of energy for local communities. The continued evolution of alternative turbine technology, digitally enabling remote monitoring, and the introduction of hybrid renewable energies are all extending the reach of small hydropower.

In addition to the continued evolution of both hydropower generation methods, there are several ways to view the different paths being taken by the various regions. In Europe and North America, emphasis is on modernizing, cleaning up the environment, and developing pumped-storage projects to assist intermittent renewable sources. Developing countries are continuing to build large-scale hydropower facilities, although environmental and political challenges will determine how far and how quickly these projects are completed. The Middle East is investing in water-energy facilities that support multiple uses and will help alleviate the devastating effects of water shortages. The development of hydropower capacity is expanding rapidly in Southeast Asia, particularly along the Mekong River, and is working to mitigate ecological and social impacts. Far East Asia, particularly China, continues to be the global leader in the hydropower sector and is developing both conventional and pumped-storage installations.

Development of hydropower in global scale

Hydropower is one of the leading renewable energy resources in the world today, as it generates a major portion of clean electricity (about 14%–16% of total power generation) and underpins the national energy systems (electric grids) for many countries globally. While there are many benefits to hydropower such as being a source

Hydropower Plants. DOI: https://doi.org/10.1016/B978-0-12-824520-0.00005-8

of carbon-free electric power, providing flexibility in the electric grid and managing water resources, increasing energy security as well as economic development, the most significant benefit of hydropower continues to be its ability to support efforts being made to combat climate change and to facilitate the transition from fossil fuels to renewable energy sources by allowing investment and innovation to be made at a large and small scale.

At the end of 2024, installed hydropower capacity worldwide grew to nearly 1450 GW, continuing its rapid growth despite fluctuations from year to year. The country with the largest amount of installed hydropower is China, which accounted for more than a quarter of total installed capacity at approximately 421 GW, followed by Brazil (~110 GW), the USA (~102 GW), and Canada (~84 GW), which combined represent a significant portion of the global hydroelectric fleet. The majority of new hydropower capacity added across all regions in 2024 occurred in the East Asia and Pacific region, particularly in China; other significant national additions occurred in Africa and South and Central Asia. The continued growth of hydropower capacity in Europe and the Americas is comparatively modest compared to previous years. The global hydropower pipeline is currently estimated to exceed 1075 GW, including a large share of pumped storage hydropower (PSH) projects, which provide an increase in energy storage capacity along with traditional hydropower generation (Enerdata, 2025; IHA, 2025).

Regional capacity highlights

East Asia and the Asia-Pacific region saw a rapid increase in electricity generation capacity from hydropower, largely due to the construction of 14.4 GW of hydropower facilities in China during 2024, which included large pumped storage projects (Hydropower.org, 2025).

In Africa, approximately 4.5 GW of hydropower capacity was put into operation in 2024, an increase of more than 100% over recent regional growth and the first time in many years that many regions have begun to develop their potential for virtually unlimited demand for hydropower (Hydropower.org, 2025). In

In South-Central Asia, approximately 4 GW of hydropower capacity was put into operation during 2024, signaling continued growth in countries like Bhutan and Pakistan (Enerdata, 2025).

Both Europe and the Americas saw small increases in hydropower capacity in 2024, but hydropower is still a critical component of both regions' ability to meet their renewable power generation goals and to implement additional battery storage systems (Enerdata, 2025).

These regional patterns reflect both resource endowment and policy frameworks. For example, China's expansive hydropower strategy includes ambitious **pumped-storage targets** to support grid stability as variable wind and solar generation increase (Reuters 2025).

Small-scale, microgrid, and hybrid developments

Although massive hydropower projects remain vital to the electrical infrastructure of nations, the growth of small hydropower plants, local grids, and storage or hybrid renewable technologies demonstrates how improving resilience and decentralizing electrical generation can also provide reliable and affordable renewable energy sources for local consumers. Small and micro-hydropower systems, from kilowatts to several megawatts, are widely used in rural communities as well as to supply power to isolated areas and for the creation of local communities' microgrids, since the design allows for local resource use and requires less capital and less environmental impact than traditional power plants. (MG 2026)

Localized energy system (Microgrid Knowledge, 2016): a microgrid is a localized electrical supply and distribution network that can operate independently or in conjunction with the larger electrical grid, and it is becoming increasingly popular for integrating renewable energy sources. Examples of hydropower integration with other renewables include integrating existing and newer renewable technologies into microgrids. The use of hybrid systems in microgrid configurations has increased the stability of energy delivery to remote areas and off-grid locations through the combination of different renewable resources with energy storage technologies, which smooths out the variability of renewable generation due to its intermittent nature (Dosa et al., 2025; Jurasz et al., 2018; Ren et al., 2022). Additionally, utility-scale hybrid hydropower systems are being developed, including hybrid hydro/solar systems, using existing hydrotransmission systems as a base to efficiently utilize existing infrastructure and increase renewable generation (Wikipedia, 2026).

With distributed/hybrid energy solutions, hydropower adaptability is demonstrated across an evolving landscape as the industry recognizes the distinction between traditional large-scale installations and a more flexible, resilient approach to delivering rural power, as well as integrating renewables and building nature-based, resilient infrastructure.

Large-scale development

Large-scale hydropower developments with generating capacities greater than 100 MW are used for most of the world's electricity generation and these developments continue to be important sources of power for many nations. The majority of such developments take place within those countries where electrical power consumption has increased rapidly and where governments desire to develop electrical energy sources that provide reliable and permanent energy solutions.

Three Gorges Dam (China), Belo Monte Dam (Brazil), and Grand Renaissance Dam (Ethiopia) are the most famous examples of "mega-dams" built to create or expand the global hydropower industry. Although these mega-dams generate very large amounts of electricity, they are also associated with many adverse effects, including (1) environmental degradation, (2) displacement of large numbers of people from their homes, and (3) increasing geopolitical tensions among nations that share common water resources. Therefore, the majority of governments and international organizations working in the area of hydropower are trying to develop and commercialize, utilizing more sustainable environmental practices (including deployment of fish migration systems, sediment management practices, and conservation strategies for ecosystems).

Small-scale development

Small-scale hydropower is defined as projects with a capacity of 10 MW or less. There is an increased global interest in small-scale hydropower because it is a decentralized and environmentally friendly energy source. These types of small-scale hydropower projects are essential for rural and remote communities where extending the electrical grid is not an economic option. Additionally, mini-hydro- and microhydroprojects (defined as projects with an installed capacity of 1 MW or less) provide a dependable source of electricity to remote or isolated communities, which allows those communities to develop economically and reduces their need for fossil fuels.

The advancement of technology in turbine design, digital monitoring, and the combination of hybrid systems with solar and wind energy will make small-scale hydropower more viable as a source of electricity. For example, Nepal, Bhutan, and Kenya are three countries that have successfully implemented mini-grid hydropower systems and have proven that sustainable and community-based forms of energy can be developed.

Development by regions

Europe

In Europe today, there exists a robust hydropower production industry that many of the countries have adopted; examples include countries such as Norway, Switzerland, and Austria, which have relied upon hydropower as their primary source of electricity for decades. As technology continues to evolve within Europe, it appears now that the trend is moving away from constructing large-scale hydropower plants or developing large-scale new developments, to focusing on modernizing and optimizing the current infrastructure in place.

The EU (European Union) has also placed an emphasis on how the hydropower sector will evolve by way of promoting "Pumped Storage Hydropower" as a way in which to connect the fluctuating renewable energy sources (i.e., wind and solar) into the grid. In addition, due to environmental issues and strict regulatory requirements surrounding the sustainability of projects, there is a much larger focus on sustainable developments in hydropower projects today.

Many of these sustainability initiatives, such as retrofitting existing dams to include fish-friendly turbine technology, enhancing the management of sediment, and decommissioning old hydropower plants, which caused disruption to ecosystems, have been driven by the EU Green Deal and Hydropower Sustainability Standard.

Europe has been active within the hydropower sector for a number of years and, through the renovation and rehabilitation of its existing hydropower plants, has continued to incorporate modern developments based on sustainability best practices into its construction of new hydropower plants. Presently, about 36% of all electricity generated from renewable resources in Europe comes from Hydropower, with Norway, Sweden, France, and Austria as the top four nations producing the most hydropower in Europe. Further are the key trends in European hydropower development:

- Modernization and refurbishment of existing hydropower plants—there is a trend for many existing hydropower plants to upgrade their equipment with high-efficiency turbines, better fish passage options, and add digital monitoring technology.

- The growing demand for pumped storage hydropower-pumped storage hydropower (PSH) is also seeing renewed investments. Pumped storage hydropower is important in maintaining grid stability as more intermittent renewable energy, such as wind and solar, comes online. Good examples are recent - all above 1000MW projects in development in Slovakia, Serbia, Romania, Spain, Greece, and Poland.
- Environmental sustainability in Europe: European environmental policies place an emphasis on the ecological balance. Therefore, projects that comply with the EU Water Framework Directive and biodiversity conservation efforts will promote environmental sustainability.

Norway is the leader in hydropower production in terms of providing the majority of its electricity via hydroelectric facilities, with approximately 90% of all electricity in Norway produced this way. Switzerland and Austria have recently expanded the use of pumped storage to create an energy storage capability to support their surrounding regions' electricity needs. In addition, some Eastern European nations, namely Albania and Bosnia and Herzegovina, are currently working to increase their production of small-scale hydropower facilities.

North America

North America and, especially, Canada and the United States have already developed a very large amount of hydropower capacity. About 60% of Canada's electricity comes from hydropower (mostly in Quebec and British Columbia). In the United States, most of the river systems are already fully developed so much interest has been placed on upgrading existing dams rather than building new ones.

The modernization of hydropower in North America is being driven by digital technology, integrative storage systems, and environmental concerns. The Columbia River Basin in the United States and Canada will continue to be a key area to balance energy production with environmental sustainability, especially as it pertains to salmon runs.

South America

Hydropower development is growing quickly in South American countries as a result of significant capital expenditures by Brazil, Chile, and Argentina on new large hydrodevelopment projects. The Amazon Basin continues to grow rapidly in hydropower generation capacity; yet the expansion of hydropower is causing significant concerns about deforestation, biodiversity loss, and potential violations of Indigenous peoples' rights, all of which have led to continued discussions about possible solutions to sustainable development in South America.

The development of hydropower in Colombia and Peru has largely been through the use of "run-of-river" technologies, which aim to minimize impacts on the environment during development. In contrast, Brazil continues to expand its extensive hydropower network so that it can meet the increasing demand for electricity. Therefore, both political and economic stability are likely to have a significant effect on the future direction of hydropower development projects throughout South America.

Africa

Hydropower potential in Africa is also massive and underdeveloped, especially along the Nile, Congo and Zambezi rivers. The construction of the Grand Renaissance Dam (GERD) in Ethiopia is a major project that can greatly increase energy access throughout the region. Nevertheless, the ongoing geopolitical tensions with downstream countries, like Egypt and Sudan, show that transboundary water management is multifaceted.

In addition, many rural African communities are now beginning to use small-scale hydropower systems as a means of gaining access to electricity, something they would not be able to attain otherwise. Countries like Rwanda and Kenya have begun developing mini-hydropower grids to achieve sustainable means of electrifying their rural areas. External funding and partnerships will help to unlock Africa's entire hydropower potential.

Middle East

Due to a lack of fresh water and dry conditions, the Middle East has few opportunities for generating energy from hydropower. Many Middle Eastern countries are developing hydropower facilities on their biggest rivers; for example, Iran and Turkey have built a number of large-scale hydropower plants. Turkey has a project called The Southeastern Anatolia Project (GAP), which plays an important role in Turkey's electricity and food production.

As the region faces increasing challenges from global warming and rising temperatures that lead to decreasing supplies of dam water (or dammed rivers), there is now a need for multi-purpose dams (the term used for dams that generate electricity and also manage river flows) throughout the Middle East, as these will help to manage water supplies better. There is also a need to investigate new methods of providing fresh water, such as desalination, and other forms of clean energy, including hybrid renewable energy systems, to augment the limited hydropower resources in the Middle East.

South, Central, and East Asia Region

The hydropower development in South-East Asia is large, driven by increasing global energy consumption and the many available water sources. As one of the most affected countries in the world, Laos, known as the "Battery of South East Asia," has aggressively developed its hydropower sector to export its energy to Thailand and Vietnam creating a Regional Power Grids—Laos-Thailand-Malaysia-Singapore Power Integration Project (LTMS-PIP) where Singapore began importing up to 100 MW of hydropower in 2022, with plans to expand up to 200 MW or Sarawak Import Project approved to import 1 GW of low-carbon electricity from Sarawak, Malaysia with expected commencement in 2035 via 700 km of sub-sea cables. Additionally, Bhutan developed over 2300 MW of installed capacity with potential for 30 000 MW. Hydropower currently supplies 99.7% of Bhutan's energy mix, and energy selling mainly to India accounts for over 40% of Bhutan's national revenue.

On the other hand, the rapid pace of hydropower development in the Mekong River Basin has raised critical concerns about environmental and social impacts. From a hydropower perspective, the confluence of the Mekong and Tonle Sap Rivers in the heart of Phnom Penh (Cambodia) has been identified as a site with significant hydropower potential, as the Tonle Sap River reverses direction between the rainy and dry seasons, acting as a natural water pump. However, issues such as flooding, diminishing fish stocks, and the displacement of local communities historically bound to riverine life are highly controversial. Regional cooperation initiatives, such as those spearheaded by the Mekong River Commission, have developed programs to help create an ecological balance between development and sustainability.

Far East Asia

Far East Asia, particularly China, leads the world in hydropower development, boasting the largest installed capacity, the largest dam, the largest hydropower plant with the largest installed capacity or the largest hydropower cascade with the highest installed capacity and, finally, the largest installed capacity per unit (per runner—1 GW [1000 MW]—single turbine). The country continues to expand its hydropower network, including pumped storage projects to complement its growing solar and wind capacity, along with a combination as cogeneration units (pumped storage and solar farm).

Japan, South Korea, and Taiwan, on the other hand, are focusing on modernizing existing facilities and investing in small-scale hydropower. The target solutions are prioritized. The integration of hydropower with smart grid technology and energy storage to enhance grid stability and reliability is part of energy policy agendas.

Conclusion

As global demand for energy increases, hydropower development is rapidly evolving to meet these demands and at the same time address environmental/social objectives. In emerging markets, particularly, large-scale hydroelectric projects will continue to provide essential sources of stable, low-cost electricity and, therefore are the foundation for continued economic development. Conversely, there is an increasing trend for small and microhydropower generation to provide flexible, decentralized solutions for rural communities, remote locations, and distributed energy networks. In addition to providing an equitable source of renewable energy, smaller facilities also increase the accessibility of renewable energy to more rural areas and reduce the environmental impacts from larger dams and reservoirs that are commonly associated with large-scale hydroelectric facilities.

Across regions, hydropower developments show varying strategies based on resource availability, local conditions, and governmental policy priorities. Developed countries emphasize modernizing the efficiency of their existing plants to continue using them as a resource—they are performing refurbishments on their aging facilities, creating a network with Smart Grid technologies, and using environmentally conscious design standards to reduce the ecological impacts of their facilities. In contrast, developing economies are constructing new facilities to meet their rapidly increasing demands for power from newly electrified areas that are taking advantage of the low prices and renewable nature of hydropower to facilitate their economic growth and eventual industrialization and urbanization. Regardless of region, there is an increased focus on maintaining a balance between energy production and the ecological system, whether it's through creating fish passage solutions, managing sediments, or protecting riverine biodiversity.

Globally, the future of hydropower will be affected by continuing technological change, continuing changes in governmental policies, and an increasing number of countries working together. Innovative turbine designs and developments in the digital monitoring area are creating opportunities for improved efficiency and reliability, as are the use of hybrid technologies—such as solar or wind power combined with hydropower—to implement more reliable, sustainable, and ecofriendly solutions. Government policies that are being used worldwide to promote renewable energy and sell or price carbon have also had an impact on the way that investment is made, as well as the way that operating practices are implemented. The current focus on achieving net-zero greenhouse gas emissions and meeting worldwide climate targets results in hydropower remaining an essential source of renewable power, by providing constant, adaptable, and low-carbon solutions, while moving toward using the most appropriate, environmentally sound, and socially accountable practices.

List of abbreviations

EU European Union
MW Megawatt
PSH Pumped Storage Hydropower

References

Dosa, A., Olanrewaju, O. A., & Mora-Camino, F. (2025). A comprehensive review of hybrid renewable microgrids: key design parameters, optimization techniques, and the role of demand response in enhancing system flexibility. *Energies, 18*(19), 5154. https://doi.org/10.3390/en18195154.

Enerdata. (2025). Over 24 GW of hydropower capacity added worldwide in 2024. Retrieved from https://www.enerdata. net/publications/daily-energy-news/over-24-gw-hydropower-capacity-added-worldwide-2024.html

Hydropower.org. (2025). *2025 World Hydropower Outlook: Global growth, trends & net zero progress.* International Hydropower Association. https://www.hydropower.org/publications/2025-world-hydropower-outlook.

IHA. (2025). *2025 World Hydropower Outlook: Global growth, trends & net zero progress.* International Hydropower Association. Retrieved December 22, 2025, from https://www.hydropower.org/publications/2025-world-hydropower-outlook.

Jurasz, J., Mikulik, J., Krzywda, M., Ciapała, B., & Janowski, M. (2018). Integrating a wind- and solar-powered hybrid to the power system by coupling it with a hydroelectric power station with pumping installation. *Energy, 144*(C), 549–563. https://doi.org/10.1016/j.energy.2017.12.011.

Microgrid Knowledge. (2016). Integrating run-of-the-river hydropower into a microgrid for reliability. https://www.microgridknowledge.com/editors-choice/article/11432602/integrating-run-of-the-river-hydropower-into-a-microgrid-for-reliability

MG. (2026). Small hydropower market size, share, growth, and industry analysis, by type (electromechanical equipment, infrastructure), by application (small hydro (1MW-10MW), mini hydro (100kW-1MW), micro hydro (5kW-100kW)), regional insights and forecast to 2035. Retrieved December 22, 2025, from https://www.marketgrowthreports.com/market-reports/small-hydropower-market-111026?utm

Reuters. (2025). China on track to exceed 2030 pumped storage hydro target by 8%, industry body says. Retrieved December 15, 2025, from https://www.reuters.com/sustainability/climate-energy/china-track-exceed-2030-pumped-storage-hydro-target-by-8-industry-body-says-2025-06-24/?utm_source%20

Ren, Y., Ren, L., Zhang, K., Liu, D., Yao, X., & Li, H. (2022). Research on the operational strategy of the hybrid wind/PV/small-hydropower/facility-agriculture system based on a microgrid. *Energies, 15*(7). https://doi.org/10.3390/en15072466 Article 2466.

Wikipedia. (2026). Bui hydro-solar hybrid project. In *Wikipedia, The Free Encyclopedia*. https://en.wikipedia.org/wiki/Bui_Hydro%E2%80%93Solar_Hybrid_Project.

7

Legislative

This chapter provides an extensive summary of the international rules governing hydropower plants and illustrates how these regulations determine facility location, development, operation, and decommissioning across various geographic areas. The main purpose of hydropower legislation globally is to establish an equilibrium between energy production and the preservation of nature, along with social obligations and economic efficiency. In general, hydropower plants are categorized by their capacity to generate electricity and the extent of their environmental consequences. All hydropower regulations contain similar components, including water rights, licensing, environmental impact assessments, and grid integration.

Many countries use laws and international agreements to guide the management of their water resources, including the Convention on the Protection and Use of Transboundary Watercourses and the United Nations Framework Convention on Climate Change. This chapter provides an overview of the differing hydropower-related regulatory incentives across regions (e.g., the EU, the United States, China, and India).

It also discusses common themes found in hydropower across different regions of the world, for example, the need for environmental impact assessments, protection of indigenous and local communities, and adherence to stringent safety standards. The chapter also provides a comparison of environmental regulations for hydropower across Europe, North America, South America, Africa, and Australia. Overall, it emphasizes the need for a solid understanding of the legislative environment for hydropower to develop compliant, sustainable, and socially responsible hydropower globally.

Introduction

Hydropower has been a major part of the world's energy history since the beginning of civilization and continues to be one of the most popular and utilized sources of renewable energy. In 2024, hydropower provided approximately 47%–48% of all renewable sources of electricity produced worldwide and was approximately

Hydropower Plants. DOI: https://doi.org/10.1016/B978-0-12-824520-0.00002-2

14% to 15% of the total production of all electricity produced worldwide (International Renewable Energy Agency IRENA 2025; International Energy Agency [IEA], 2025). Hence, it provides a significant amount of renewable electricity being produced globally and is critical to energy security and stability of the grid (International Energy Agency [IEA], 2023). Hydropower plants use the natural motion of water to produce reliable, low-carbon-emission electricity for decades or more. In addition, most hydropower developments provide additional public benefits such as flood control, irrigation, and potable water supply and provide ancillary service support to power networks, making them vital elements within both national energy systems and regional energy development strategies (World Bank, 2018; International Hydropower Association [IHA], 2020).

As a renewable energy source, hydropower is fundamentally linked to natural and cultural resources. Building/managing/improving/dismantling a hydroelectric facility impacts on not just the generation of electricity, but also the ecological system of the area and the use of land by the surrounding community. The impacts on the ecology and use of land and water brought about by hydropower development include the alteration of river flow patterns, rearrangement of sediment movement, interruption of fish movement between locations, and forced relocation of communities resulting from the creation of large storage reservoirs (World Commission on Dams [WCD], 2000). To account for these many impacts, hydropower projects are subject to complex legal/regulatory systems that differ from country to country. Each of these rules is designed to balance the need to generate electricity with the need to protect the environment, consider the social impacts of the project, and ensure its long-term economic *viability* (UN Environment Programme [UNEP], 2017).

Internationally, hydropower development is dictated by international agreements and funding obligations and sustainability guidelines that support the responsible planning and operation of hydropower projects. These frameworks provide guidance to user nations on best practices for producing and operating a hydropower project, including the need for an Environmental Impact Assessment and the inclusion of the affected public in project development, as well as the protection of biodiversity and the development of risk management strategies for large projects, particularly those that involve more than one nation (United Nations, 1997; World Bank, 2018). While these principles are essentially the same across the globe and provide a consistent path forward, the ability to implement them often depends on the local legal systems, governance structures, and local capacity to do so.

Across the region, there is a broad range of regulatory rules applying to hydropower; some jurisdictions regulate hydropower under their energy policies while others regulate it primarily under their water resource management/environmental protections/land-use planning regulations (IEA, 2023). In addition to varying by jurisdiction, each regulatory approach is tailored to meet specific hydropower project types, such as large reservoir dams; run-of-the-river projects; pumped storage; or small or micro hydropower systems. In several countries, many ageing hydropower facilities are presenting valuable opportunities for regulatory authorities to enhance regulations regarding safety upgrades, environmental retrofits and licensing processes (IHA, 2020).

The goal of this chapter is to present a comprehensive examination of how hydropower is governed through legal regulations globally and regionally. The chapter begins with an overview of some of the key global frameworks impacting the development and financing of hydropower, and then will explore some of the regional approaches taken in Europe, North America, Asia, Africa, and Australia. A comparison of the various perspectives presented in this chapter shows that although the legal systems that govern development of hydropower differ, they all share the same challenge of developing hydropower in an inclusive, responsible and sustainable manner.

Global hydropower legislation

General definition of Hydropower plants by law on a global scale

Regulatory Law relating to Hydropower Plants is enacted by Laws in each country of the World. Although the specific laws may differ from Country to country, many of them share similar guiding principles in the area of Hydropower Development and the same Legal Definition for the Classification of Hydropower Projects (IHA, 2020). Legal Definitions in general provide a means to distinguish between Hydropower Development classifications based on Instantiated Capacity, Environmental Impact, and Ownership Type (IHA, 2020).

To make the Legislation simpler for the purpose of regulation of hydropower development, the majority of existing Legislations provide for categorization or classification of hydropower Development, generally, between three basic categories, namely, large-scale hydropower projects—generally with a capacity greater than 10 MW, small-scale hydropower projects—generally between 1 MW and 10 MW, and microhydropower projects (generally less than 1 MW), although, categorizations and classifications vary by regulatory authority (UNEP, 2017).

Apart from classification, the following aspects of hydropower development are governed by laws associated with hydropower plants, water rights and allocation, licensing and concession regimes, environmental protection, dam safety standards, connection to transmission and market participation (World Bank, 2018), not only to ensure development of Hydropower Projects that are technically, economically and environmentally and socially sustainable. At the global level, there are many international organizations that promote the sustainable development of hydropower. Examples of these organizations are the International Energy Agency (IEA) and the United Nations (UN). Through research and international cooperation, these agencies provide a policy framework to assist in developing hydropower projects. Policy guidelines for hydropower development provided by these agencies also require that projects be developed in compliance with environmental and social safeguard measures while encouraging the growth of renewable energy resources (IEA, 2023; United Nations, 2015). Thus, many countries require that hydropower developers complete an Environmental Impact Assessment (EIA), have public consultation, and obtain a formal permit before they can begin constructing their hydropower project (WCD, 2000).

Small hydropower plants

The definitions of small hydroelectric facilities vary by country, but for the majority of cases, small hydroelectric facilities are defined by their installed capacity, usually between 1 and 30 MW; however, a few countries have an upper limit of 50 MW (IHA, 2020). The regulatory provisions that apply to small hydroelectric projects generally differ from the regulatory requirements that apply to larger projects; specifically, small hydroelectric facilities tend to have less complex licensing processes than larger projects, while also being subject to fewer environmental assessments and less stringent grid access specifications.

The European Commission has defined a small hydro power plant in Europe as one with a generating capacity of up to 10 MW (European Commission, 2020). In the United States, there is no formal single national definition of small hydro power plants, but many authorities consider facilities operating with an installed capacity between 5 and 30 MW as small hydro (Federal Energy Regulatory Commission [FERC], 2022). China and India use a definition of small hydro as anything with a nameplate capacity of up to 50 MW, which indicates the importance of rural electrification and decentralized energy production for these countries (International Renewable Energy Agency [IRENA], 2019). Some governments provide incentives to promote small hydro power through mechanisms such as feed-in tariffs, preferential access to the electrical grid, or subsidies, acknowledging the role that small hydro power can play in meeting renewable energy goals and its reduced negative effect on the environment as compared to larger storage dams (IRENA, 2019).

International treaties and agreements

Several international agreements and organizations directly or indirectly influence hydropower legislation worldwide. Some of these include:

1. **The Kyoto Protocol and Paris Agreement**—These agreements set emission reduction targets and promote renewable energy, including hydropower (United Nations, 2015).
2. **United Nations Watercourses Convention (1997)**—This treaty provides legal principles for the use and conservation of transboundary water resources, impacting international hydropower projects (United Nations, 1997).
3. **International Hydropower Association (IHA) Guidelines**—While not legally binding, these guidelines offer best practices for sustainable hydropower development (IHA, 2020).
4. **World Bank and IFC Performance Standards**—These financial institutions impose environmental and social requirements on hydropower projects seeking international funding (World Bank, 2018).

Common regulatory themes

Despite differences in national legislation, several key themes emerge globally:

- **Environmental Protection**—Laws requiring environmental impact assessments (EIAs) to mitigate impacts on biodiversity and ecosystems (WCD, 2000).
- **Water Rights and Usage**—Regulations governing the allocation and use of water resources for hydropower (UNEP, 2017).
- **Social and Indigenous Rights**—Laws ensuring the protection of local communities, including compensation and resettlement requirements (World Bank, 2018).
- **Safety and Technical Standards**—Regulations mandating dam safety, maintenance, and operational guidelines (IHA, 2020).

Regional hydropower legislation

Europe

The European regulatory framework for hydropower development represents the highest level of commitment to environmental and energy regulations within a country. The EU places a strong emphasis on the environment, sustainable water use and the integration of climate policy, and as such, hydropower projects in Europe are regulated at both the EU and national level through multiple layers of legislation, which often require coordination among the authorities that regulate energy, environment, and water resources.

The goal of European hydropower regulation is to find a balance between meeting renewable energy targets, and protecting river ecosystems, biodiversity, and water quality. This has become increasingly important as many European rivers are heavily regulated, and the existing hydropower infrastructure is aging. Regulatory attention now focuses on modernizing existing plants, developing environmental mitigation measures, and relicensing existing plants rather than new large-scale dam construction.

The key components of the regulatory framework consist of the EU-directives that establish binding targets for environmental protection and energy use, and the National Laws that govern how and when hydropower projects are licensed, operated, and meet environmental standards. Below is a list of the key components of the regulatory framework at both the EU and National levels:

1. **European Union (EU) Directives:**
   - o **Water Framework Directive (WFD)**—Requires member states to maintain or improve water quality, influencing hydropower licensing and operation.
   - o **Renewable Energy Directive (RED II)**—Sets renewable energy targets, indirectly encouraging hydropower investment.
   - o **Habitats and Birds Directives**—Protects biodiversity and restricts hydropower development in ecologically sensitive areas.
2. **National Regulations:**
   - o **Germany**—The Renewable Energy Act (EEG) promotes hydropower but imposes stringent environmental standards.
   - o **France**—Laws require extensive environmental assessments and community consultations for new hydropower projects.
   - o **Norway**—As a hydropower leader, Norway has a strong regulatory framework, including the Water Resources Act and licensing by the Norwegian Water Resources and Energy Directorate (NVE).

America

The hydropower industry in North America is well regulated through established legislative structures. These frameworks include the environmental oversight, coordination between federal and state/provincial governments, and the long-standing tradition of regulating this industry. Each country's hydropower laws are intended to create a balance among producing clean and renewable energy from hydropower while at the same time managing water resources and preserving the environment and public interest.

In the United States, federal regulation of hydropower is based on the Federal Power Act (FPA). This law gives the Federal Energy Regulatory Commission (FERC) the authority to license and regulate most non-federal hydropower projects. The licensing process requires FERC to administer the requirements of other federal laws that affect hydropower projects, including National Environmental Policy Act (NEPA), Clean Water Act, Endangered Species Act and many others, so that environmental impacts and stakeholder concerns are addressed through the entire life of a hydropower project.

Provinces in Canada primarily regulate hydropower, since they are responsible for controlling and managing natural resources. However, the federal government does have laws that apply to hydropower projects when they affect fish habitat, or indigenous rights involving navigable waterways and lands within the respective jurisdiction(s). The laws previously described, when combined with the relevant provincial laws, create a strong legal foundation for sustainable hydropower development within North America. Below is a summary of all of the various federal regulations that apply in regards to hydropower projects in both the United States and Canada.

North America

1. **United States**
   - o **Federal Energy Regulatory Commission (FERC)**—Issues licenses for hydropower plants and enforces environmental regulations.
   - o **Clean Water Act (CWA)**—Governs water quality standards that impact hydropower operations.
   - o **National Environmental Policy Act (NEPA)**—Requires EIAs for hydropower projects.
   - o **Endangered Species Act (ESA)**—Limits projects affecting protected species and habitats.
2. **Canada**
   - o **Fisheries Act**—Protects fish habitats, impacting hydropower projects.

- o **Provincial Regulations**—Each province, such as British Columbia and Quebec, has specific hydropower laws and permitting processes.

Latin America

Latin America relies extensively upon hydropower as a source of electricity. Hydropower has the potential to provide the region's national energy security as well as be part of the energy trading market among countries in the region. Hydropower's regulation varies widely between countries, as do the countries' policies on renewable energy resources and environmental protection. For instance, under Brazil's National Water Agency (ANA) water use permits are issued and enforced by this agency. However, IBAMA (Instituto Brasileiro do Meio Ambiente e dos Recursos Naturais Renováveis) enforces IBAMA's environmental regulations, which include environmental impact assessments and social safeguards. In Chile, projects for hydropower must comply with the SEIA (Sistema de Evaluación de Impacto Ambiental) to ensure that the projects satisfy environmental, and community, requirements. Also in Argentina, the Water Code and national energy laws provide the framework for licensing, allocation of water, and operational requirements of hydropower projects. Therefore, within Latin America, laws and regulations governing hydropower projects typically integrate environmental impacts, social engagement, and technical specifications, thus illustrating the commitment of the region to sustainable management of natural resources and renewable energy development. The regulations below illustrate selected examples.

- **Brazil**—The National Water Agency (ANA) oversees water use permits, while IBAMA enforces environmental laws.
- **Chile**—Hydropower projects must comply with the Environmental Impact Assessment System (SEIA).
- **Argentina**—The Water Code and federal energy laws regulate hydropower licensing and operation.

Asia

The regulatory frameworks for hydropower are highly diverse across Asia. This diversity reflects the many differences within Asia in terms of political systems, environmental priorities, and different levels of economic development. Water management, environmental protection, and energy policy differ across Asia, resulting in different approaches to planning, licensing, and operating hydropower projects in each country. In China, the leading hydropower developer across Asia, large-scale hydropower projects, such as the

Three Gorges Dam, undergo a very extensive environmental impact assessment process and meet regulatory requirements for water management. Moreover, investment in hydropower is promoted through the Renewable Energy Law as a part of China's overall strategy for climate change and energy security. The regulatory framework in India for hydropower development allows for private investment into hydropower projects and requires compliance with the Environmental Protection Act and Forest Conservation Act in particular when the hydropower project is located in an ecologically sensitive area. In Japan, Thailand, Vietnam, and other Southeast Asian countries, a combination of national energy laws and international financing requirements governs the development of hydropower projects, with many countries cooperating through regional organizations such as the Mekong River Commission. The diverse regulatory frameworks across Asia illustrate the importance of understanding local legal and regulatory frameworks in order to develop hydropower sustainably and responsibly. The list below summarizes the major hydropower players in Asia, including their respective regulatory requirements.

1. **China**
   - The **Renewable Energy Law** promotes hydropower, but projects must adhere to strict EIAs and water management regulations.
   - The **Three Gorges Dam** and other major projects have faced criticism, leading to reforms in social and environmental policies.
2. **India**
   - The **Electricity Act (2003)** encourages private investment in hydropower.
   - The **Environmental Protection Act** and **Forest Conservation Act** regulate projects in ecologically sensitive areas.
3. **Japan**
   - Hydropower is highly regulated. It is governed by the **Electricity Business Act** and the **Rivers Act**, ensuring sustainability and safety. **Environmental Impact Assessment (EIA)** mitigates potential environmental and social impacts. Large-scale projects require environmental impact assessments (EIAs) under the national Environmental Impact Assessment Law (EIA Law). The small projects with installed capacity ranging from 200 to 1000 kW are handled at the municipal level (Japan Hydropower Market Trends & Growth Report 2030, 2023; Wu et al., 2023).
4. **Southeast Asia**
   - Countries like Thailand, Vietnam, Cambodia, and Indonesia have hydropower policies influenced by international funding requirements.
   - The **Mekong River Commission** plays a role in regional hydropower coordination.

5. Africa

In Africa, since only about 10% of the continent's technical potential has been developed (International Hydropower Association, 2022), many African nations are increasing their capacity for hydropower to provide electricity as the demand for electricity continues to grow, provide more energy security to their citizens, and encourage sustainable economic development. Most African hydropower projects receive funding or, in many cases, partial funding from international development banks, private foreign investors, and multilateral investment agencies. Such financing will allow these projects to raise the large amount of capital needed to complete construction and operation. On the other hand, target financing is vague, and most of the projects, despite securing approval, are not yet under construction. In fact, the African continent is facing many challenges, but the most common ones include stalled project progress at this very stage. The 2024 IHA coordinated development Abuja Action Plan addresses potential hindrances in hydropower project delivery (International Hydropower Association, 2022). Hydropower projects, which are governed by a combination of national, regional, and international regulations, are designed to achieve a balance between energy development and environmental and social responsibility, as well as the economic benefits of energy use. The main legal structures that apply to these developments are the African Union's energy policies, which promote renewable energy development across the African continent, and the Nile Basin Initiative and Southern Africa Power Pool Frameworks, which promote cooperation among multiple countries in water management for hydropower development. At the national level of regulation, a number of countries, including South Africa, Ethiopia, and the Democratic Republic of Congo, have specific national legislation to regulate water usage, obtain licenses, conduct environmental impact assessments, and engage with the communities affected by the development of hydropower projects, to ensure that the development of hydropower is effective and sustainable. The following is an abbreviated list of international, regional, and national regulations or agreements that apply to African hydropower projects:

1. **Regional and International Agreements:**
   - o The **African Union (AU) Energy Policies** promote renewable energy, including hydropower.
   - o The **Nile Basin Initiative** and **Southern African Power Pool** coordinate hydropower policies.
   - o The **Abuja Action Plan**—a high-level strategic framework referring most notably to sustainable hydropower development in Africa. *(Initiated by IHA)*

2. **National Regulations:**
   - **South Africa**—the National Water Act governs hydropower development.
   - **Ethiopia**—hydropower expansion is central to energy policy, regulated by the Ethiopian Electric Power Corporation.
   - **Democratic Republic of Congo**—the Inga Dam projects follow national and international environmental guidelines.

Australia

Environmental considerations and national energy policies greatly influence the hydropower sector in Australia. The majority of Australia's hydropower resources are located in a limited number of regions, particularly in the south-east states, including Tasmania, where rainfall patterns and river systems support both reservoir and run-of-river schemes. To be able to develop hydropower sites, developers have to comply with environmental regulations under the Environmental Protection and Biodiversity Conservation (EPBC) Act. The EPBC sets out requirements for developers to assess and mitigate their impacts on ecological systems, 12 endangered species, and river health prior to project approval. Additionally, Water Management Law (Water Act 2007) establishes the framework for how the waters are allocated and used; therefore, when developing hydropower sites, all necessary measures must be implemented to ensure that the ecological flow of the river and of other uses dependent upon the water resources are not compromised. Coupled with this are the national renewable energy targets (renewable energy target [RET]), which provide a financial incentive to investment in hydropower in addition to investment in other forms of renewable energy (wind and solar), due to both hydropower's ability to provide reliable, low-carbon electricity and its role in balancing electricity supply with demand through providing stability for the electricity grid. Finally, licensing and permitting processes are established at the state level, reflecting both state-specific ecological priorities and sensitivities toward local environments, and act as another layer of regulation. Therefore, the development of hydropower in Australia is an example of how the clean production of energy can be achieved in conjunction with sound environmental stewardship and sustainable water management. A comprehensive (5380 files and RES headings) list of the applicable laws is listed below:

1. **Commonwealth and state laws**
   - Environmental Protection and Biodiversity Conservation (EPBC) Act—requires approval for projects impacting biodiversity.
   - Water Act (2007)—regulates water use, affecting hydropower projects.

- o State Regulations—each state, such as New South Wales and Tasmania, has specific hydropower licensing rules.

2. **Renewable energy targets**
   - o The Australian Renewable Energy Target (RET) incentivizes hydropower as part of clean energy goals.

Conclusion

Given the wide array of factors influencing hydropower legislation, including environmental, social, economic, and political aspects, it is naturally complex and differs from region to region. Ecological sensitivity, availability of water, the way river basins are managed, demand for energy, and institutional capacity are some of the differences that determine how hydropower is regulated at both the national and regional levels (IEA, 2023). Therefore, the framework of laws governing hydropower development will typically reflect local priorities, whether those priorities are based on the protection of the environment, energy security, economic growth, or social equity.

The establishment of a uniform approach to the sustainable development of hydropower within the context of international agreements and guidelines is essential. Establishing guidelines, such as Environmental Impact Assessments, Managing Transbounded Waters, Preparing and Adaptive Management of Dam Emergency Studies, and Developing Guidelines for Stakeholder Participation, is the main objective of International Agreements and Guidelines in the area of hydropower development (United Nations, 1997; World Bank, 2018). Despite the fact that there are International Frameworks for Hydropower Development, National Policies will always supersede these Global Frameworks for Hydropower Development. As a result, any International Policy or Agreement will be interpreted at the national level based on each country's own government structure and development priorities, which will lead to the implementation of international standards being adapted to each country's governmental structures and development goals. This multilayered approach to hydropower development presents unique challenges and opportunities, especially for internationally funded hydropower developments, cross border river systems, and developers with multiple nationalities.

To help stakeholders of international hydropower developers, operators, regulators, financiers, and communities affected—their ability to understand the associated laws is vital. Meeting regulatory compliance not only decreases the potential liability or financial impact but also increases transparency and better decision-making, as well as greater environmental protection and improved results socially. As hydropower is developed in response to changing conditions due to climate change, energy transition, and aging infrastructure, so too shall its legal framework change. By developing legal and regulatory alignment and coherence among state jurisdictions, establishing best practices, and supporting hydropower legislation that meets long-term sustainability goals, hydropower can continue to play an effective and sustainable role in our evolving global energy mix.

List of abbreviations

ANA	National Water Agency (Brazil)
AU	African Union
CWA	Clean Water Act (United States)
EEG	Renewable Energy Act (Germany)
EIA	Environmental Impact Assessment (Japan)
EIA Law	The National Environmental Impact Assessment Law (Japan)
EIAs	environmental impact assessments (Japan)
EPBC	Environmental Protection and Biodiversity Conservation Act (Australia)
ESA	Endangered Species Act (United States)
EU	European Union
FERC	Federal Energy Regulatory Commission (United States)
IBAMA	Brazilian Institute of Environment and Renewable Natural Resources
IEA	International Energy Agency
IFC	International Finance Corporation
IHA	International Hydropower Association
MW	Megawatt
NEPA	National Environmental Policy Act (United States)
NVE	Norwegian Water Resources and Energy Directorate
RET	Renewable Energy Target (Australia)
RED II	Renewable Energy Directive II (European Union)
SEIA	Environmental Impact Assessment System (Chile)
WFD	Water Framework Directive (European Union)

References

European Commission. (2020). *EU guidance on hydropower and the Water Framework Directive*. European Union.

Federal Energy Regulatory Commission. (2022). *Hydropower licensing and regulation*. FERC.

International Hydropower Association. (2022) *Africa hydropower regional profile. Hydropower in Africa*. Retrieved December 20, 2025, https://www.hydropower.org/region-profiles/africa.

International Energy Agency. (2023). *Hydropower special market report*. IEA.

IEA (2025), Global Energy Review 2025, IEA, Paris https://www.iea.org/reports/global-energy-review-2025, Licence: CC BY 4.0.

International Hydropower Association. (2020). *Hydropower sustainability assessment protocol*. IHA.

International Renewable Energy Agency. (2019). *Renewable energy policies for small hydropower*. IRENA.

International Renewable Energy Agency IRENA 2025.

Japan Hydropower Market Trends & Growth Report 2030. (2023). Markets and Data. Retrieved December 20, 2025, https://www.marketsandata.com/industry-reports/japan-hydropower-market.
United Nations. (1997). *Convention on the law of the non-navigational uses of international watercourses.* UN.
United Nations. (2015). *Paris Agreement.* UNFCCC.
UN Environment Programme. (2017). *Water and energy governance frameworks.* UNEP.
World Bank. (2018). *Environmental and social framework.* World Bank.
World Commission on Dams. (2000). *Dams and development: A new framework for decision-making.* Earthscan.
Wu, K.-C., Lin, J.-C., Chang, W.-T., Yen, C.-S., & Fu, H.-J. (2023). Research and analysis of promotional policies for small hydropower generation in Taiwan. *Energies, 16*(13), 4882. https://doi.org/10.3390/en16134882.

8

Environmental impact

The environmental impact

Globally acknowledged as a clean, mature, and, from a historical point of view, the first commercially applied renewable energy source, hydropower generates electricity by harvesting energy via conversion of the kinetic and potential energy of the flowing or falling water with the help of hydraulic structures such as accumulation and diversion structures, or simply through a run-of-river configuration where the impact of hydraulic structures is kept to the minimum.

As mentioned previously, hydropower, being one of the oldest and widely established renewable energy technologies, plays a major role in reducing global dependence on fossil fuels, along with mitigating greenhouse gas emissions, and if it comes to the sole energy generation, hydropower is a low-carbon footprint technology, but on the contrary, despite historically argued that the hydropower generation is not a subject of consumptive water user rather than a nonconsumptive water user, hence keeps the low water footprint index, but if it comes to evaporation ration of reservoirs this statement could be indeed put into a serious challenge (Mekonnen & Hoekstra 2011).

Nevertheless, despite its wide range of advantages, hydropower—physically perceived through hydraulic structures equipped with mechanical and electrical internal technology—is related to civil, mechanical, electrical engineering, and architecture; therefore, the impact of its development can have significant environmental and social consequences if not designed with a sense of environmental impact and responsibility. The impacts of hydropower vary considerably depending on plant size, installed capacity, design, geographic location, and cumulative development within a river basin. This chapter examines the positive and negative environmental impacts of hydropower plants, with particular attention to the differences between large-scale and small hydropower installations, and introduces the mitigation strategies necessary for sustainable development.

Hydropower Plants. DOI: https://doi.org/10.1016/B978-0-12-824520-0.00010-1

Positive environmental contributions of hydropower in general

Hydropower is a very important form of electric power generation that can provide a positive benefit toward environmentally oriented sustainable development, while many of the environmentally adverse claims about hydropower are based on exaggerated information and poorly communicated intentions by non-engineering-based groups. Hydropower has significantly reduced greenhouse gas emissions compared to other fossil-fuel electric power production (i.e., coal, oil, natural gas). The amount of Greenhouse gases produced by hydropower generation equipment is substantially less than that of fossil fuel-based power plants, as hydroelectric plants actually do not directly produce carbon dioxide or any other harmful air pollutants (e.g., SO_2, NO_x, particulate matter, CO, or HPAs) in the process of generating electricity.

This means that hydropower generation helps reduce the negative effects of climate change, enhances air quality for people living in communities around hydropower plants, and decreases potential adverse health effects (AHE) caused by air pollutants. Hydropower-producing countries with a high percentage of hydropower in their total electricity generation (e.g. Brazil, Canada, Norway, Paraguay) generally have lower CO_2 emissions on an electric energy consumption basis than those that do not (Fig. 8.1). Hydropower is an effective low carbon energy resource, and since water as a part of small and large water cycle, is renewable, thus it can produce electricity for as long as there are sufficient hydrological conditions available. The oldest hydropower plants have lifespans of over a century; for example, the 1918 Slovenian Fala Hydropower plant has been operating for over 100 years, and the even older Chinese Shilongba Hydropower plant has been active since 1912 (Ghosh, 2022). The modern hydropower turbine technology is cost-competitive form of electricity as the initial investment is relatively high (depending on the size, type, and location) (Student Energy, 2015), but the efficiency ranges from 85% to 96% (Andritz, 2026), and they can last from 25 up to 80 (Blume-Werry & Everts, 2022), but reaching also 100 years (Hydropower, 2004) with minimal additional investment, making it a reliable, efficient, and economically long-lasting one-time investment compared to other renewable energy systems. What is more, hydropower is stable and pairs well with other sources, so it can be used to reinforce the base-load power typically dominated by fossil- or nuclear-powered plants (Fig. 8.2).

In recent years, hydropower has been increasingly viewed as a trusted source of energy and a dependable partner to renewable

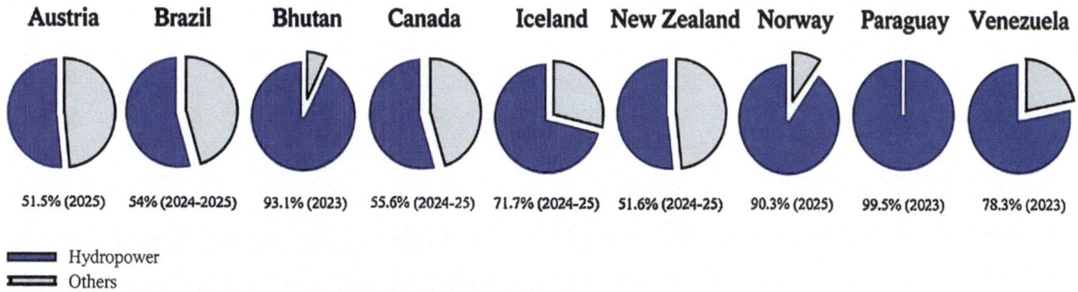

Figure 8.1 Energy generation from hydropower in the various countries where the hydropower sector is leading the energy industry (representing a significant share of the energy mix).
Author's personal drawing based on https://lowcarbonpower.org/.

energy sources to create Hybrid Renewable Energy Systems (HRES). The reliability of HRES has made it a vital element in developing long-term energy security and in the broader transition of energy systems toward sustainability and resilience. An HRES can also be implemented on its own, in isolated or remote areas, or as a backup for an existing conventional electricity network, creating a more robust and reliable electricity generation system.

Beyond energy production, hydropower infrastructure can offer additional environmental and socioeconomic benefits by creating reservoirs. These reservoirs are seldom built for a single function and are almost exclusively designed as long-term, sometimes even strategic (e.g., a source of drinking water for the entire region or cross-border flood protection), multipurpose hydraulic structures. Common reservoir functions may include supplying water for irrigation canals (agricultural function), flood protection, drinking/utility/industrial water storage (water consumption function), and cooling industrial processes (coolant function). Many areas provide recreational activities and support a range of water sports, fisheries, and tourism to promote local economic development. Reservoirs can also serve as fish farms within the biodiversity-food nexus.

Overall, hydropower is a multipurpose, environmentally beneficial element of renewable energy strategies that will benefit current and future generations alike.

Negative environmental contributions of hydropower in general

While hydropower is widely regarded as a reliable and sustainable source of energy due to its renewable nature and low carbon

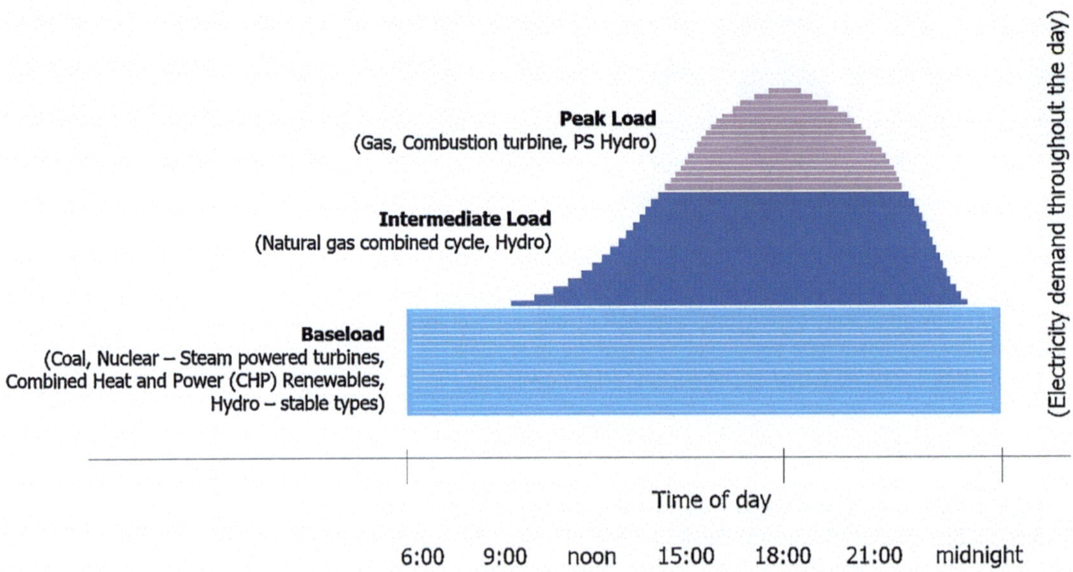

Figure 8.2 Schematic explanation of the hydropower incorporated into the daily energy demand scheme: base load, intermediate load, and peak load.
Author's personal drawing.

footprint, it also entails significant ecological, environmental, and social disadvantages that must be considered during planning and construction. Before digging deeper, it is important to separate the impact of hydropower during construction from its sole energy generation. These two cannot be separated when discussing the overall environmental impact of hydropower, as one represents the initial impact at the beginning of the life span and the other the impact during operation, thus the lifespan of a powerhouse. Now, regarding the circular economy, these two are also inseparable, as the real benefits of the hydropower plants are in the length of the operation lifespan, and in the case of urban renewal or functional retrofitting, the powerhouse structure can continue to exist with numerous alternative functions, such as a restaurant or a museum.

One of the greatest impacts of hydropower facilities is the potential disruption of natural ecosystems caused by the construction of diversion and retention structures, such as dams and reservoirs. This also includes temporary impediments, such as coffer dams, during construction. In general, the diversion or retention structures fundamentally alter the natural flow patterns of rivers, including tributaries. They also create barriers in river systems, preventing fish and other aquatic organisms from migrating along their natural routes and interfering with the life cycles of species that depend on longitudinal connections between river systems. On the other hand, reservoirs can in fact act as a breeding ground for fish farms, hydropower turbines oxygenate water, which is seen as a positive impact; hence, it is important to look for a balance between the water body's existing fauna and flora, and how the hydropower project can be of help, rather than pose a disruption. In addition, dams disrupt natural sediment transport processes, leading to sediment accumulation in reservoirs and sediment starvation downstream, which, in turn, affects river morphology, delta stability, and the health of downstream habitats. Closely linked to these changes is biodiversity loss, as reservoirs often inundate extensive areas of terrestrial ecosystems, leading to the permanent loss of habitats for plants and animals. Transformations from a dynamic Riverine environment to a static lacustrine environment support limited species diversity; many species (particularly endemic and endangered species) cannot adapt and will thus decline or become extinct. Erosion of water quality in Hydropower Reservoirs is an additional consequence of their creation, as impoundment causes stratification, dissolved oxygen depletion, and changes in the thermal regime. These factors contribute to increased rates of eutrophication and algal blooms, while the decomposition (or breakdown) of this organic material produces Methane, a powerful greenhouse gas. Beyond environmental concerns, large hydropower projects often entail profound social consequences, including the displacement of local populations to make way for reservoirs. In general, large-scale reservoir projects require the complete demolition of areas to be flooded, but in practice, many villages, cities (China, Italy, Spain, Greece), or even old hydropower plants (Japan) have been submerged in their original state, some of which have become tourist attractions during seasons when the water level is low and structures start appearing from the water. On the other hand, in some cases, such as the ancient Han Dynasty city of Shicheng in China, submerged in Qingdao Lake in 1950 (DwellScape, 2024), a submerged city perfectly preserved over time raises questions about the entire submersion process, in contrast to heritage preservation, as very little needs to be done to keep the city intact while attracting divers and media attention. For most cases, relocation as a result of intentional flooding can be detrimental to those affected by it regarding their way of life, society, and culture; these effects are even more pronounced for indigenous peoples' (or Native Inhabitants of the Land) ability to remain connected with their ancestral lands. In addition, the impoundment (i.e., "storage") of huge quantities of water comes with inherent risks associated with both seismic and geological activity; specifically, reservoirs have an increased risk of causing earthquakes (i.e., "Reservoir-Induced Seismicity" [RIS]) for those living near dams in areas subject to tectonic movement.

Despite well-established designs, the Buttress and Embankment (also known as Earth/Rockfill Dams), as well as adjustable barriers such as adjustable barrages or movable weirs, are among the most susceptible to failure. On the contrary, all dam structures, whether directly or indirectly connected to hydropower, located in tectonic areas are subject to permanent monitoring for mass shifts or structural failures.

Environmental impacts of large hydropower plants

Although large hydropower facilities have many advantages, they also pose serious environmental and social problems. For example, when large dams and reservoirs are built, they fundamentally change the way rivers function naturally; flooding a very large area can cause microclimates (climate zones) to develop, affecting the characteristics not only of the local fish populations, but if large enough, it may alter the flora and fauna of the region around it. Hence, the rule of thumb is that the larger the reservoir to be constructed, the wider the area needs to be studied for potential environmental impacts. Additionally, river flow regulation has caused sediment transport to stop downstream, thereby altering seasonal river flows and creating fragmented aquatic habitats. As a result of these issues, downstream fisheries, river morphology, and delta erosion have been negatively affected, leading to numerous changes across the entire watershed.

Biodiversity loss is a serious issue. Because large reservoirs flood large areas, they effectively destroy large portions of terrestrial ecosystems for plants and animals. When moving rivers to reservoirs, the species that will survive will often be a very select few, while species adapted to live in moving water will be removed or displaced. Migratory fish species are among the first to feel the impact of dams, which restrict access to breeding and gene-flow points along natural migration routes.

Water quality can also deteriorate in large reservoirs. Thermal stratification, reduced oxygen levels, and nutrient accumulation may promote eutrophication and harmful algal blooms. Furthermore, the decomposition of submerged organic matter—especially in tropical regions—can release methane, a potent greenhouse gas, partially offsetting the climate benefits of hydropower.

The social effects of large hydroelectric projects can be significant. Creating reservoirs may require relocating entire communities, resulting in the loss of livelihoods, cultural heritage, and social networks. Indigenous people are often most affected and struggle to replace their economic (including spiritual), cultural, and social connections to land and rivers. Additionally, dams are susceptible to earthquakes and typhoons. What is more, the large volumes of water behind the dam can trigger earthquakes, a phenomenon called "reservoir-induced seismicity" that poses a threat to public safety and infrastructure in areas with tectonic activity. Seismic activity and its impact on modern dam structures are constantly evaluated using numerous sensors; modern dam designs are relatively safe. The bigger problem is typhoons and the resulting alluvial deposits, which can significantly shorten reservoir lifespans by decades, thereby jeopardizing large investments; thus, additional desilting and dredging are usually set up.

In general, large-scale hydropower plants have a range of environmental impacts, some of which are perceived as serious hazards. Gigantic projects such as the Three Gorges Dam, a technological and civil engineering challenge and still perceived as the largest hydropower development in the world, also carry an array of environmental impacts (Hvistendahl, 2024), which eventually reinforced the Energy Decentralization and Small Hydropower Movement. On the contrary, large-scale development is possible, and if the environmental impact is properly studied, it might result in an environmentally acceptable and economically viable design. Although the current trend in large-scale hydro is slowly shifting toward pumped storage units, which are usually built in caverns and, although visually unobtrusive, are in fact underground structures, they are subject to complex boring engineering using Tunnel Boring Machines (TBMs).

Environmental impact of small hydropower plants

Small hydropower plants (SHPs; i.e., electric plants with a capacity of 10 megawatts [MW] Maximum) (IEA Hydropower, 2024) are commonly championed as a green alternative to large-scale Hydropower Facilities. SHPs' small resource requirements, low upfront costs, and ability to generate electricity in a distributed manner for a long period of time make them especially useful for the electrification of rural areas and other remote and isolated locations.

That said, SHPs (excluding submersible Run-of-River (RORs) systems) are also hydraulic structures and require site development;

therefore, they are subject to environmental impact regulation. Even modest amounts of water diverted from their natural flow path will likely affect how rivers transport sediments, cycle nutrients, and create habitat for fish and other aquatic animals. While ROR-type SHPs do not require the construction of large storage tanks, they may reduce the amount of water passed through a section of river directly downstream of an installation (especially in the summertime). This creates heightened stress on aquatic species that rely on continuous flow to migrate and successfully reproduce. This is commonly addressed through biocorridors and fish passages, some of which (such as Brush blocks Fish Passage) mimic natural conditions (Kucukali et al., 2023). A common problem with biocorridors or fish passages is navigating fish toward the entrance, which can be tricky in some cases, and a more advanced solution, such as a low-voltage barrier, is necessary. What is more, recently, some small hydropower turbine producers have promoted fish-friendly turbines, claiming successful fish survival rates over 98%. Taking into account turbine pressure, blades, and overall system economics, such turbine solutions become questionable, particularly regarding the level at which a given fish species can withstand rapid pressure changes. In general, the larger the installed capacity, the more serious the problem of fish survivability.

Despite significant technological and landscape developments, the issue of fish migration remains a significant concern for the development of SHPs; thus, where possible, biocorridors are advised. Still, many SHPs lack an effective passage for fish to migrate past the installation, thereby preventing their movement and their ability to successfully reproduce in the area above the facility. Though an individual SHP may have localized impacts on the fish migration process, the combined total of numerous SHPs installed on a single river can generate substantial fragmentation of fish habitats, resulting in a decrease in fish species diversity.

SHPs generally have smaller environmental disruption but can still affect sediment transport. The presence of Low Head Weirs can cause sediment build-up behind them and lead to downstream erosion, riverbed incision, and habitat degradation in some cases or areas, especially in mountainous areas or when sediment transport is affected by erosion. Water quality may also be adversely affected as low flow velocities create higher surface temperatures and reduced oxygen levels, and increase the concentration of any pollutants in the water column.

Additionally, the impact of land use resulting from an SHP must be considered. The construction of ancillary items (access roads, transmission lines, etc.) is likely to cause deforestation, soil erosion, and habitat fragmentation, especially if built in or adjacent to ecologically sensitive or designated areas. Climate change introduces additional uncertainty about the amount of water available due to changes in precipitation, patterns, and seasonal variability. These changes will create more ecological stress and increased competition for water between users and make it more difficult to develop an effective management strategy for an SHP.

Conclusion

Overall, hydropower is an important source of renewable energy that can provide many benefits, including low greenhouse gas emissions, reliable electric power, energy recovery, and many other products and services, such as water supply, irrigation, and flood control. While large and small hydroelectric facilities both have the potential to create substantial negative environmental and social impacts, including habitat destruction, diminished biodiversity, changes to river flows, and the displacement of local communities, small hydropower plants typically have a smaller physical footprint than large hydropower projects, but, if aggregated in a hydropower cascade, they can have a cumulative impact which might be in some measures even worse in comparison to large scale projects.

However, given the cumulative effects of multiple smaller hydropower plants on local ecosystems, it is still vital to evaluate these factors before constructing new hydroelectric facilities. Sustainable development of hydropower requires a thoughtful balance between the positive impacts of maximizing energy production for society (installed capacity and annual energy delivery) and minimizing negative ecological impacts (the environmental impacts of each gigawatt of installed capacity). Many successful approaches to building sustainable hydropower projects include mitigation measures such as providing fish passes and biocorridors, using fish-friendly turbine technology, managing sediment transport, and maintaining the natural flow of water downstream of hydroelectric dams.

List of abbreviations

AHE	Adverse Health Effects
CO_2	Carbon Dioxide
HPAs	Hazardous Air Pollutants
HRES	Hybrid Renewable Energy Systems (HRES)
MW	Megawatt
NO_x	Nitrogen Oxides
PM	Particulate Matter
RIS	Reservoir-Induced Seismicity
SHP	Small Hydropower Plant

SHPs Small Hydropower Plants
SO_2 Sulfur Dioxide
TBMs Tunnel Boring Machines (Moles)

References

Andritz. (2026). *Turbines for hydropower plants.* Retrieved January 23, 2026, from https://www.andritz.com/products-en/group/products/turbines-hydropower#:~:text=ANDRITZ%20Hydro%20Francis%20turbines%20are%20ideally%20suited%20for%20energy%20applications.Kaplan%20turbines.

Blume-Werry, E., & Everts, M. (2022). Hydropower. In M. Hafner, & G. Luciani (Eds.). *The Palgrave Handbook of International Energy Economics*Cham: Palgrave Macmillan. https://doi.org/10.1007/978-3-030-86884-0_8.

DwellScape. (2024). Shicheng, China: The Underwater Ancient City of Mysteries | DwellScape [Video]. YouTube. Retrieved January 23, 2026, from https://www.youtube.com/watch?v=0c42f4cHtI4.

Ghosh, A. (2022). *Multiple makings at China's first hydroelectric power station at Shilongba*, 1908–1912. *History and Technology, 38*(2–3), 167–185. https://doi.org/10.1080/07341512.2022.2112295.

Hvistendahl, M. (2024). China's Three Gorges Dam: an environmental catastrophe? Scientific American. Retrieved January 23, 2026, from https://www.scientificamerican.com/article/chinas-three-gorges-dam-disaster/#:~:text=The%20dam%20has%20been%20linked%20to%20a,the%20faults%2C%20which%20can%20lead%20to%20earthquakes.

Hydropower. (2004). *Hydropower Setting a Course for Our Energy Future [Brochure]. U.S. Department of Energy Energy Efficiency and Renewable Energy.* Retrieved January 23, 2026, from https://docs.nrel.gov/docs/fy04osti/34916.pdf.

IEA Hydropower. (2024). *What do we mean by small-scale hydro?* Retrieved January 23, 2026, from https://www.ieahydro.org/about/past-achievements-and-completed-activities/annex-2/what-is-small-hydro.

Kucukali, S., Alp, A., & Albayrak, I. (2023). *Retrofitting Vertical Slot Fish Pass with Brush Blocks*: Hydraulics Performance. *Water, 15*(6), 1155. https://doi.org/10.3390/w15061155.

Mekonnen, M. M., & Hoekstra, A. Y. (2011). *The water footprint of electricity from hydropower.* Value of Water Research Report Series No. 51, UNESCO-IHE. Retrieved January 19, 2026, from https://www.waterfootprint.org/resources/Report51-WaterFootprintHydropower.pdf.

Student Energy. (2015). *Hydropower 101* [Video]. YouTube. Retrieved January 23, 2026, from https://www.youtube.com/watch?v=q8HmRLCgDAI.

9

Dictionary of technical terms

Introduction

Hydropower is an essential renewable energy source (RES). Hydropower generates electricity by harnessing the kinetic and potential energy of water as it flows or is stored. The complexity of these systems requires that hydropower professionals have a thorough understanding of multiple engineering disciplines (mechanical, electrical, and civil) and the technical terms associated with them. This dictionary serves as a quick reference for these people, providing clear definitions of the key terms in hydropower technology and design that are currently most commonly used. Understanding the terminology will enhance hydropower engineers' and the industry's ability to collaborate effectively, improve system efficiency, and maintain compliance with established regulations and standards of practice in environmental, mechanical, electrical, and civil engineering.

Hydropower plant

A type of Civil Engineering and/or Hydraulic structure equipped with a hydropower generator to convert the kinetic and potential energy of moving water into electricity. Hydropower plants are considered renewable energy facilities, regardless of size or arrangement, which are the main factors in further sustainability and environmental assessments of the powerhouse's effectiveness (Fig. 9.1)

General typology (Fig. 9.1)

- **Terranean**
  The majority of Civil and Hydraulic structures are on the surface
  - Surface-type powerhouse
    Stand-alone powerhouse with penstock, powerhouse adjacent to an existing weir or dam structure, or a combination.
- **Subterranean**
  The Civil and Hydraulic structures are partially or fully underground

Hydropower Plants. DOI: https://doi.org/10.1016/B978-0-12-824520-0.00011-3

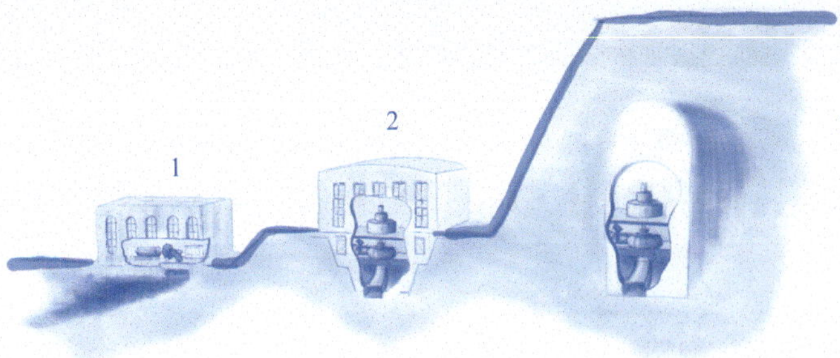

Figure 9.1 Combined picture of surface (1), semiunderground (2), and cavern-type hydropower house (3).
Author's personal drawing.

- **Semiunderground**
  A significant portion of the Civil and Hydraulic structures is underground. This configuration is ideal for a vertical machinery arrangement, with electrical equipment (generator and governing units) located in the superstructure, while the substructure houses water-related mechanical parts (turbine and accessories). In some specific cases, such as shaft installation to meet head requirements (e.g., Water Treatment Plant installations), a horizontal turbine arrangement is also possible.
- **Cavern-type**

The majority of the Civil and Hydraulic structures are underground. Exposed structures are usually diversion structures such as dams or weirs, inflow gates, surge tank shafts, service and emergency entrances, and the tailrace. Underground hydropower houses are installed in bored caverns designed to protect the hydropower equipment from harsh weather, landslides, earthquakes, and possible sabotage. The caverns are also constructed in river profiles where Terranean or semiunderground units would be harder to construct due to natural obstacles, and the sole construction would significantly reshape the original landscape. Hence, they are typically suited to strategic hydropower projects, such as Pumped Storage Units or Units with large installed capacity, where any disruption would pose a direct threat to the regional or national power grid stability.

Hydropower turbine and machinery

This section of the dictionary focuses on the basic hierarchy of water machines and their arrangement according to the position of the axis of revolution: horizontal, vertical, and inclined, respectively. Expanded typology and the most common variations are also included. However, as the turbine industry reflects rising energy demand and ecological trends, the hydropower sector is forced to adapt and bring increasingly more complex solutions for challenging scenarios, especially in the small hydropower and retrofitting sectors. Therefore, the dictionary is limited to the most common solution, current market-available technology. The dictionary also lists large-scale systems, especially pumped storage technology, that are widely affected by rising system efficiency demands. In general, the list of water machines is hierarchically sorted from the most primitive (e.g., Water wheels) to the most complex solutions (Ternary set technology).

Water wheel

- **Horizontal axis**
  The main axis of rotation is positioned horizontally toward the water level.
- **Vertical axis**
  The main axis of rotation is positioned vertically toward the water level. In some historical cases, a vertical axis could also be inclined (Device for lifting water, 2012).
- **Stream type (including floating water mills)**
  Stream-type water mills are a group of water wheels, usually with a vertical axis, that operate without any water-guiding structures, such as a flume, shaft, or forebay. The water wheel depends solely on the actual water level in the canal or river. This makes such water wheels highly susceptible to water-level changes; they can be installed efficiently only under stable conditions or with an adjustable axis height, which adds technical challenges. This was bypassed by a floating mill, in which the entire system was placed on a boat that rises or drops with the water level.
- **Overshot waterwheel**
  The water is delivered to the wheel buckets via flume or penstock over the top. Because gravity does most of the work, this solution is highly efficient, reaching up to 90%.
  - **Backshot/pitch back**
    Reversed overshot arrangement—an alternation of the overshot waterwheel, with a flume or penstock delivering water to the back of the waterwheel.
  - **Breastshot waterwheel**
    The water is delivered to the wheel buckets via flume or penstock approximately mid-height of the water wheel. Because gravity helps only halfway, this solution typically has lower efficiency than the overshot solution, usually 75%–87%.
  - **Undershot waterwheel**
    The water is delivered to the wheel buckets and the canal, but it hits only the bottom bucket. Because gravity does not help in this

matter, the solution is solely relying on the kinetic transfer between the flowing water and the bucket design. This solution is indeed the cheapest but typically has very low efficiency, reaching 50% at best.

- **Hybrid forms of waterwheel**
  Hybrid forms refer to variations and combinations of inflows that result in higher torque and/or expanding on wheel movement, allowing for turning in both directions.
  - **Overshot and backshot**
    It is a specific example where water was available at two heads, allowing the combination of both types and resulting in a higher torque ideal for forges that need to lift heavy hammers.
  - **Reversible**

The buckets are bidirectional and can operate in either direction. Ideal for hoisting mechanisms, such as mining lifts.

Impulse turbines

- **Pelton turbine (horizontal** (Fig. 9.2)**/vertical** (Fig. 9.3)**)**

Invented by American inventor ***Lester Allan Pelton (1829–1908)*** *in the 1870s, the most famous impulse turbine was based on extensive research into water mills and early impulse turbine technology. Since then, Pelton turbines have undergone various modifications, including a series of bucket designs, water-jet multiplication, vertical-to-horizontal arrangements, size variations, and dual-turbine single-shaft configurations. It is the most widely used impulse turbine worldwide. Recently, the multijet Pelton turbine became part of the Pumped Storage Ternary set technology, operating in generation mode and as a sleeve engine for the pump via a torque converter.*

- **Turgo turbine (horizontal/vertical)**
  This small impulse turbine was invented and patented by the English engineer ***Eric Crewdson (1888–1967)*** *in 1919 as a compact single bucket, single nozzle impulse turbine ideal for medium heads. Unlike the Pelton turbine, the Turgo turbine has not been as successful in large-scale installations, mainly because it is less efficient, more complex, and has blades that are more fragile in large applications. However, it is widely used in micro and pico systems with dual- or quadruple-jet modifications.*

- **Cross flow turbine (Banki–Ossberger–Michell turbine)**

It was a special impulse turbine literally developed from the water wheel, where water flows perpendicular to the shaft and through the turbine blades (transversely—hence the name crossflow). The design

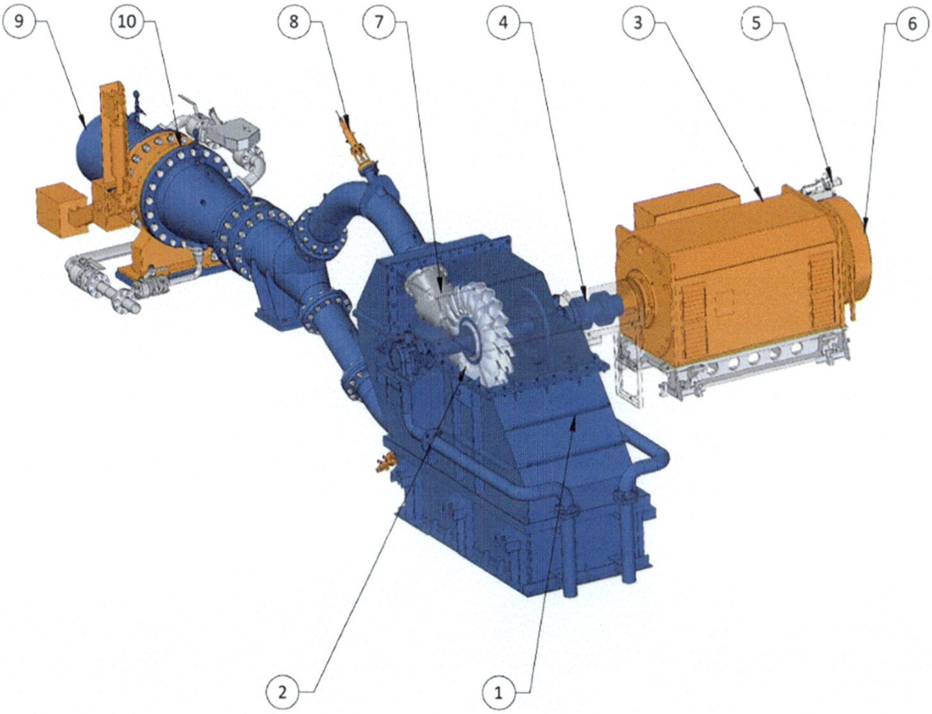

Figure 9.2 3D view of the Direct-driven Horizontal Pelton turbine and its basic parts: turbine case (1), runner (2), generator (3), shaft (4), brake (5), flywheel (6), nozzle (7), nozzle control (8), steel intake (9), butterfly valve (10). *Courtesy of Mavel a.s.*

was carried away independently by three inventors, namely, Donát Bánki (1859-1922), Fritz Ossberger (1877-1947), and Anthony George Maldon Michell (1870-1959), hence the alternative name Banki-Ossberger-Michell turbine. The turbine can have horizontal or vertical inflow and is solely regulated by the inflow guide vane. The turbine runner looks like a protruding water wheel; therefore, it is sometimes called a tubular turbine. To allow for more flow flexibility, the cross-flow turbine runners are subdivided in a 1:2 ratio and into 2 or 3 segments and could be equipped with different-sized gaps between blades on each segment. It also applies 2 or 3 separated guided vanes (distributors), allowing the turbine to be used only partially in case of low water level. The regulation allows inflow use at 33%, 66%, or 100%. The Cross-flow represents a cheaper, less efficient solution compared to the Pelton runner or Turgo, powered by more complex adjustable nozzles, but, due to its simplicity, it is a preferred turbine solution for small hydropower plants with abundant water flow (Picone et al., 2025).

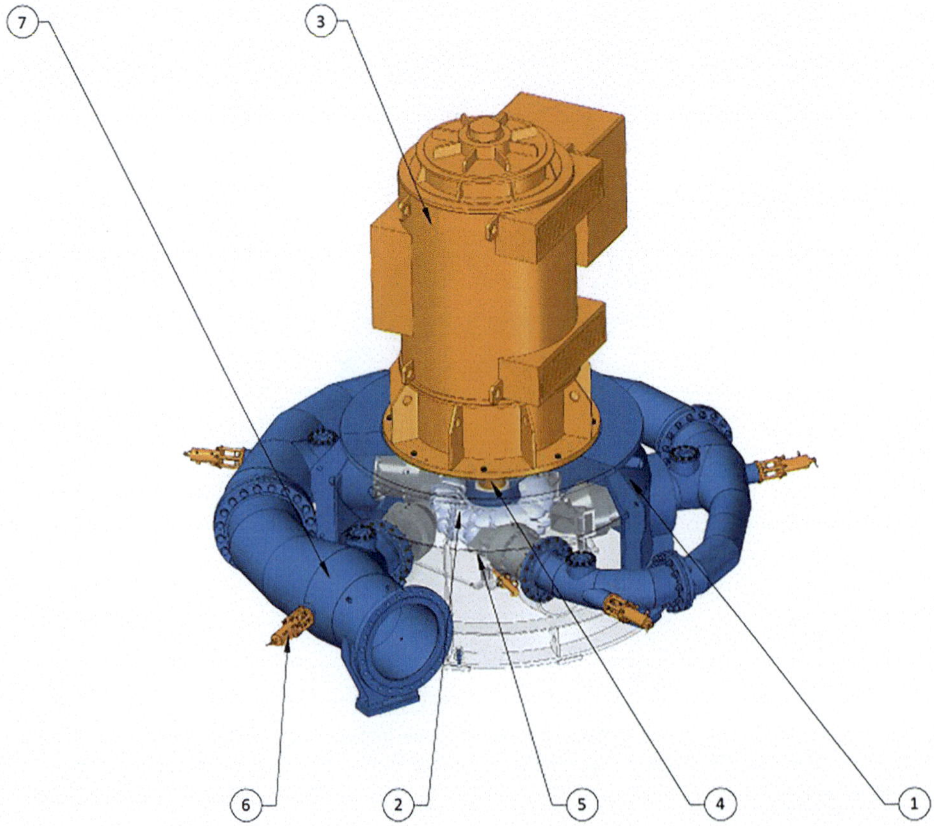

Figure 9.3 3D view of the Direct-driven Vertical Pelton turbine and its basic parts: turbine case (1), runner (2), generator with brake (3), shaft (4), nozzle (5), nozzle control (6), steel intake (7).
Courtesy of Mavel a.s.

Reaction turbines

- **Francis turbine (horizontal** [Fig. 9.4]**/vertical** [Figs. 9.5, 9.6]**)**

Due to its flexibility, robust design, and large application portfolio, the Francis turbine is the most widely used turbine technology worldwide. It was invented by British-American engineer ***James Bicheno Francis (1815–1892)*** *in 1848–1849. The turbine can operate in both horizontal and vertical configurations and has undergone multiple modifications since its original open-flume design. The most famous was the spiral casing modification, which increased the turbine's efficiency to 96%. The other modifications were double runner, camelback arrangement, or multiple module arrangement, such as dual camelback (quadruple Francis on a single shaft). Francis turbines are also commonly used as reversible turbines in pumped-storage units.*

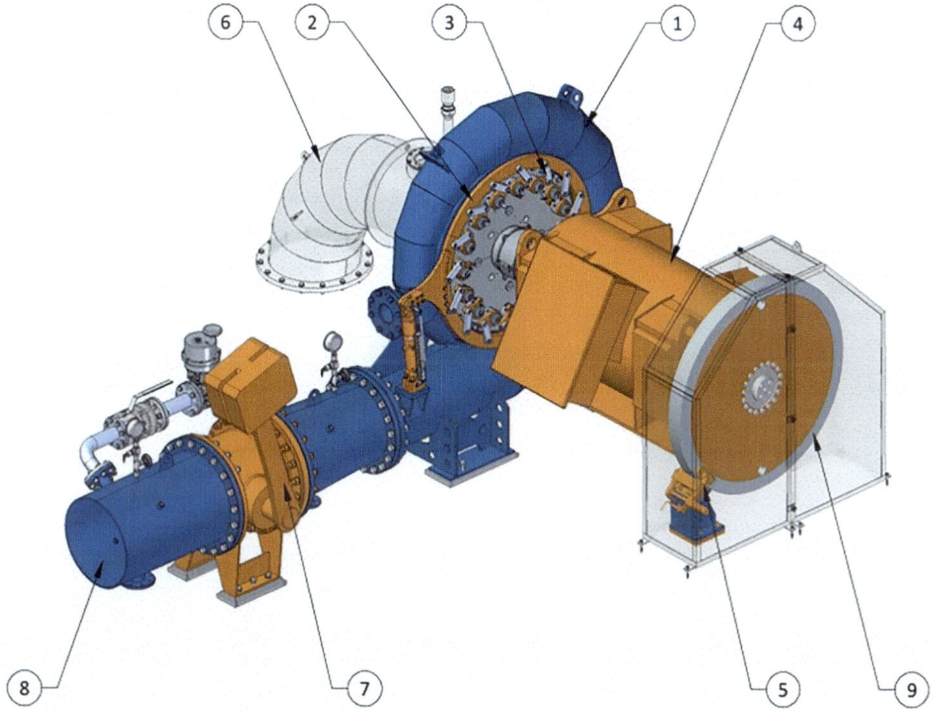

Figure 9.4 3D view of the Direct-driven Horizontal Francis turbine and its basic parts: spiral case (1), turbine distributor (2), wicket gates (3), generator (4), brake (5), draft tube (6), butterfly valve (7), steel intake (8), flywheel (9). *Courtesy of Mavel a.s.*

- **Propeller type**

  Simple ship propeller application in reversed mode with fixed blades. The Propeller turbine is a direct predecessor of Kaplan turbine technology; therefore, it is sometimes called Smi-Kaplan. The propeller turbine is also the base design for multiple other micro and pico systems, such as the Dive turbine, Vortex, Siphon, and others.

- **Kaplan turbine (horizontal** [Figs. 9.7, 9.8]/**vertical** [Fig. 9.9])
  It is invented by Austrian engineer ***Viktor Kaplan (1876–1934)*** *in 1913 as a hydropower turbine for maximum efficiency under varying load using adjustable blades. The turbine was essentially a modification of the propeller type, which has fixed blades. The Kaplan turbine became ideal for low-head installations and, since then, has undergone several modifications, including the addition of wicket gates, a dual-unit/single-shaft arrangement, a bulb* (Fig. 9.7), *a pit, and a bevel-gear turbine to a generator coupling. Together with the Francis turbine, it is the most widely used reaction technology. Under certain conditions, the Kaplan turbine could be used in a reversible mode.*

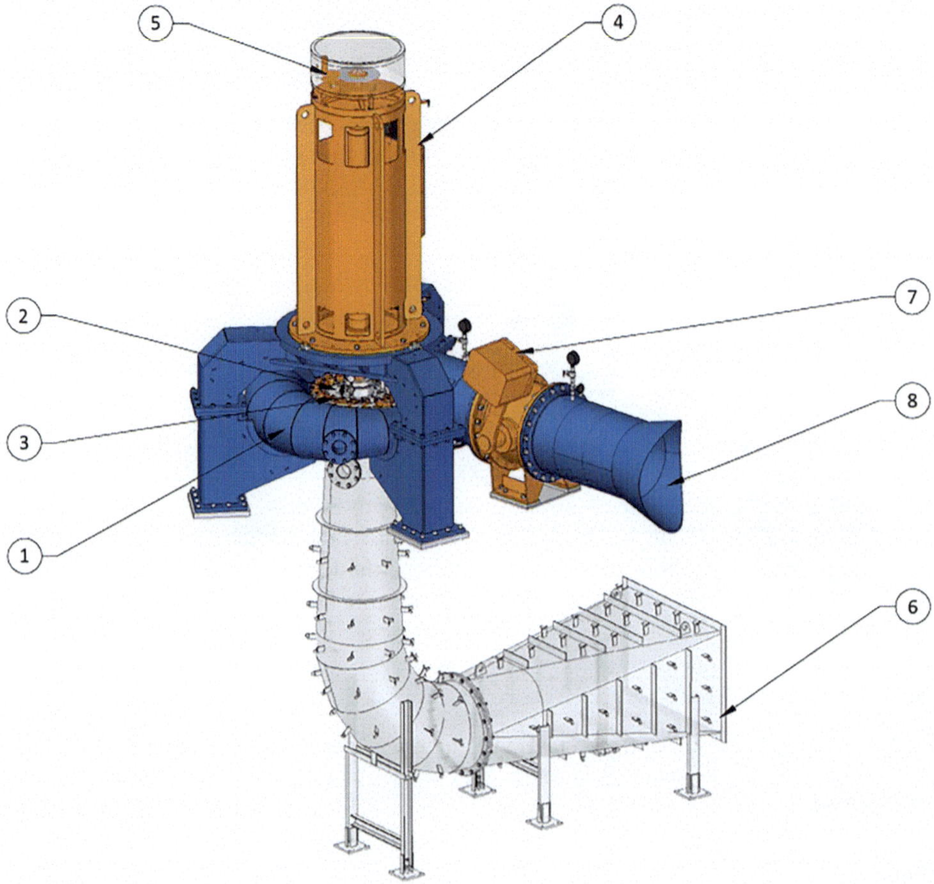

Figure 9.5 3D view of the Direct-driven Vertical Francis turbine and its basic parts: spiral case (1), turbine distributor (2), wicket gates (3), generator (4), brake (5), draft tube (6), butterfly valve (7), steel intake (8).
Courtesy of Mavel a.s.

- **Deriaz turbine**
  It is a relatively modern turbine designed by Swiss hydraulic engineer ***Paul Deriaz (1907–1990)*** *in the 50 s as a combination of a Francis turbine with Kaplan blades, capitalizing on both diagonal and axial flow, making it ideal for medium-head applications. Due to its flexible switching between pumping and generating modes, the Deriaz turbine is successfully applied in Pumped Storage hydropower plant projects as a reversible unit.*

- **Alden turbine**
  It is a relatively new type of turbine developed by ***Alden Research Laboratory****, Electric Power Research Institute (EPRI), and Voith*

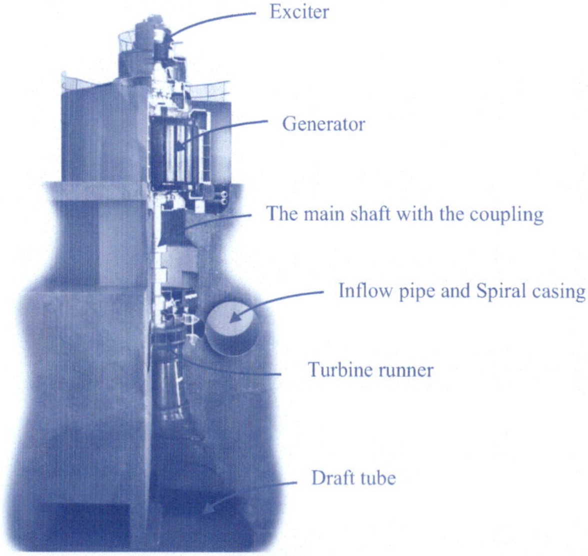

Figure 9.6 Cross-section of the Vertical-shaft Francis Turbine.
Photo by Štefan Tkáč—based on Tienlun Hydropowerplant education model, 2013.

Hydro, with a focus on environmental protection through fish-friendliness, achieving more than 98% fish survival while maintaining high efficiency. So far, only demonstration power plants have been established, and feasibility case studies have been carried out. However, Alden's turbine design, in a slight modification, has been successfully used as a base for vortex turbine technology successfully applied in the microhydropower sector.

- **Vortex turbine**
  It is a propeller-based turbine using an open spiral (vortex chamber) and a water vortex. The idea is to build on the successive improvements of the suction turbine by an Austrian inventor, Viktor Schauberger (1885–1958), and a series of free-surface vortex turbine prototypes (Timilsina et al., 2018). The vortex technology is considered an ecological, fish-friendly micro and pico solution, although in theory, the larger applications are also viable.
- **Siphon turbine**
  It is a propeller or Kaplan-based turbine that uses a suction head and negative pipeline pressure (siphon effect). The application head is limited to 6 m due to the water stream interruption. The system can be designed as modular, fully open, and exposed technology, allowing full flexibility for new applications, retrofitting, and direct weir or water treatment plant installations.
- **AST (Archimedes screw turbine)**
  Low head turbine designed as a reversed Archimedes Screw. It is widely used as a robust small hydropower unit, mostly adjacent to

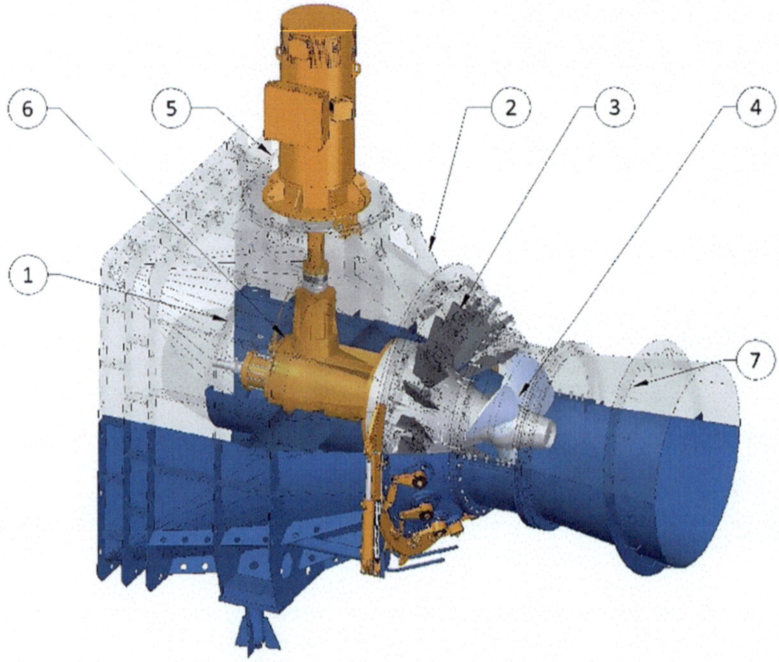

Figure 9.7 3D view of the Bevel-driven Kaplan Bulb turbine and its basic parts: turbine case (1), turbine distributor (2), wicket gates (3), runner (4), generator (5), bevel gearbox (6), draft tube (7).
Courtesy of Mavel a.s.

existing weirs or directly installed in water canals. It is availabe both in open and fully enclosed configurations.

- **DIVE turbine**

  It is a modern, compact application for small hydropower plants or even for retrofitting old water mills. It is considered a fish-friendly turbine, with a simple propeller and a submerged generator. It is constructed exclusively as a vertical-axis unit.

Medium properties turbines

The water medium's properties, such as viscosity, density, adhesion, cohesion, and high surface tension, are used in researching modern turbine technology design. As a matter of fact, the micro and pico solutions, such as the Tesla bladeless turbine or bladeless PROTUR (Precession Rolling Turbine) known as SETUR are commercially available.

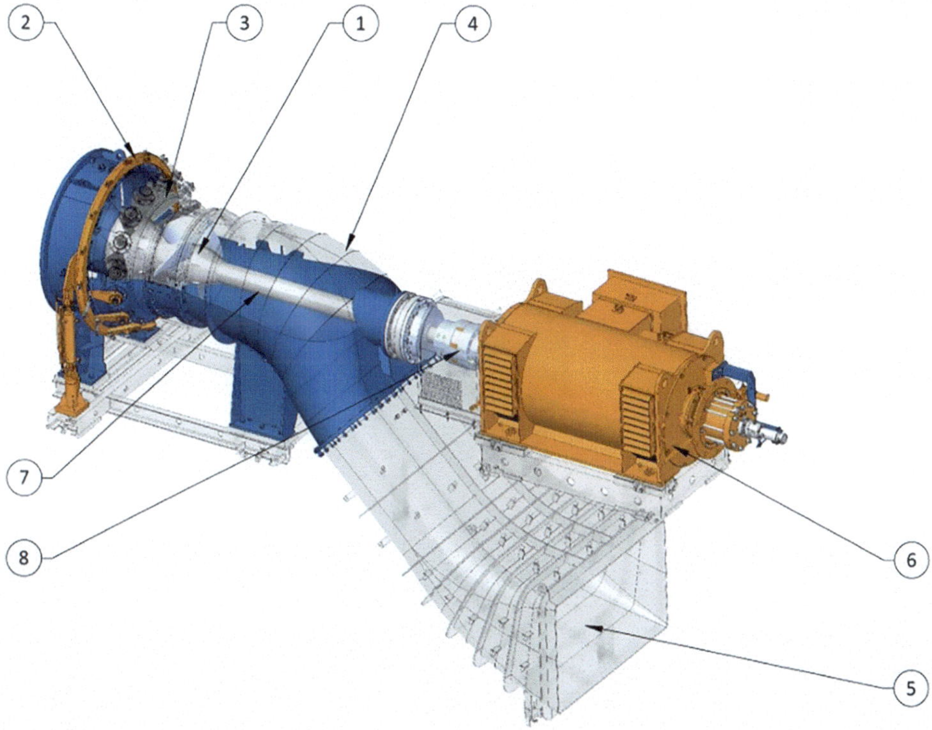

Figure 9.8 3D view of the Direct-driven Kaplan S-type turbine and its basic parts: runner (1), turbine distributor (2), wicket gates (3), turbine elbow (4), draft tube (5), generator (6), turbine shaft (7), coupling (8).
Courtesy of Mavel a.s.

Reversible turbines (pumped storage technology)

Pumped storage hydropower, or pumped hydroelectric storage, is to date one of the most proven techno-economic solutions for long-term energy storage. Pumped storage hydropower plants (PSHPP) are estimated to store between 179 and 200 GW of total global power, accounting for 90%–95% of global long-term power storage. Thus, PSHPPs are sometimes referred to as "the world's largest water battery."

- **Pump-based technology (pumped as turbines)**
  This system uses standard water pumps in pumping mode; there are examples where the pump is also used in generation mode, though with far lower efficiency, unlike reversible turbines. A multistage pump is also used in Ternary sets solely working as a pumping mechanism.

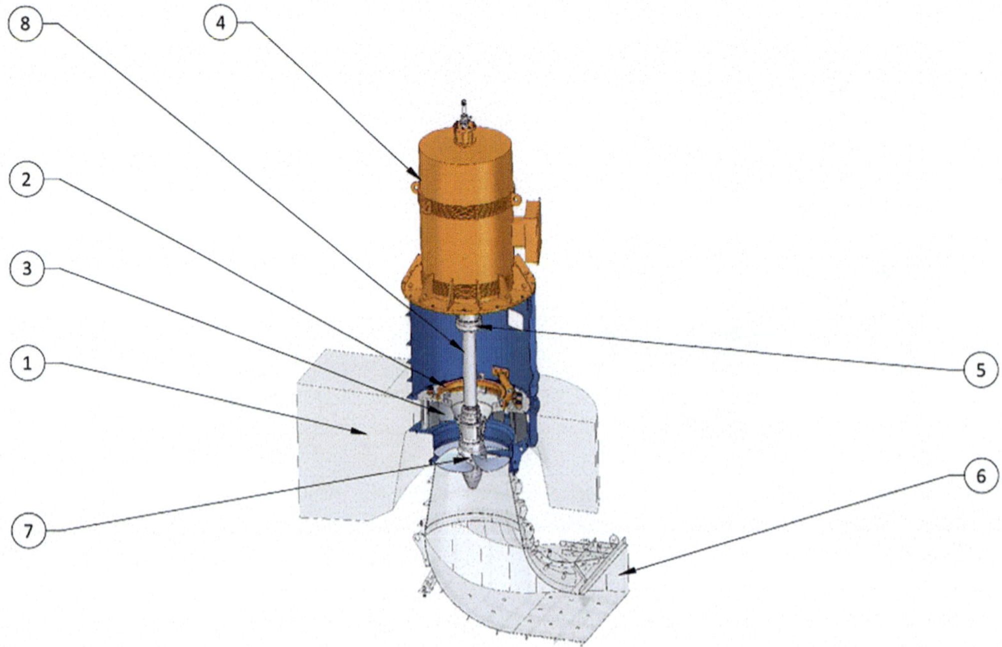

Figure 9.9 3D view of the Direct-driven Vertical Kaplan turbine and its basic parts: concrete rectangular spiral (1), turbine distributor (2), wicket gates (3), generator (4), coupling (5), draft tube (6), runner (7), turbine shaft (8). *Courtesy of Mavel a.s.*

- **Reversible turbine**

  The simple reversible turbine system is the most common and economically viable solution for pumped storage units. It consists of a generator and a reversible turbine, such as a Francis, Deriaz, or, for low heads, a Kaplan turbine. The system is less efficient when storing surplus energy from volatile renewables, and switching between pumping and generation modes takes time because the reversible turbine is doing both jobs.

  - **Francis/Deriaz**

    Both turbines are the most common reversible turbines, and Deriaz uses its adjustable blades to provide greater flexibility, especially in low-head systems. Deriaz is also a significantly newer and more complex type, designed in the 50s, while Francis was designed in the 40s. Pumped storage technology in combination with Pelton runner is also possible, for example, Grand'Maison, France, which features 4 Pelton turbines alongside 8 reversible multistage Francis pump turbines, where Pelton runner is used in high head generation mode, while Francis would work predominantly in Reversible mode.

- **Kaplan**

  The Kaplan turbine is not specifically designed as a reversible turbine and is usually low efficiency, but at low heads, the installation can have advantages due to the turbine blades' adjustability. (An example of a reversible bulb Kaplan turbine would be the Tass Water Discharging Facility, Hungary.)

- **Variable-speed Control**

  The Variable Speed Control Pumped Storage Hydropower plant (VSC PSH) offers greater flexibility in energy storage mode because the generator operates in motor mode (pumping mode) at variable speed. There are about 10 installations worldwide, as this system requires a relatively large powerhouse due to the AC excitation system and additional generator cooling. Although it allows more efficient storage of surplus electricity, including that from volatile renewables, the system still uses a reversible turbine; hence, switching between pumping and generation modes takes time.

- **Ternary set technology**

  Unlike the previously mentioned standard pumped storage (reversible) binary units, which consist of a motor-generator and a reversible turbine, the Ternary sets consist of three components, namely a motor-generator, a pump, and a turbine, all mounted on a single shaft, allowing for smoother transition between pumping and generating modes with no sleeve or pony motors. The arrangement can be vertical or horizontal (Some of the examples are Kops II (Kopswerk II) in Austria with a generator placed between the turbine and the pump, and Veytaux II in the Switzerland-France border with a generator, turbine, and pump arrangement, both are vertical sets using a Pelton turbine for power generation. In the Otherermunt II hydroelectric plant (Obervermuntwerk II), on the other hand, a horizontal Francis turbine arrangement is used, with the generator placed between the turbine and the pump.

- **Totally separated machines**

  In this arrangement, the turbine(s) and pump(s) operate completely separately, allowing for higher flexibility between pumping and generating modes. The arrangement can be vertical, as in the case of Ternary sets, or horizontal, or, in large hydraulic cascade systems, the two can be placed in separate installations: a pumping station and a turbine station.

List of common turbine abbreviations

The common turbine abbreviations are usually used for data tags or documentation materials, and they comprise information on turbine type, followed by the axis orientation, turbine runner dimension and sometimes the gearbox type and gearbox arrangement.

Turbine type

K: Kaplan turbine
F: *Francis turbine*
P: *Pelton turbine*
TT: *Turgo turbine*
CFT *(in research CFHT): cross-flow turbine (cross-flow hydraulic turbine)*
AST: *Archimedes screw turbine*
SX: *Saxo turbine*
DIVE: *DIVE turbine*
DFT: *dual Francis turbine (twin runner) usually followed by information on discharge (single or dual discharge)*

Orientation

V: *vertical*
H: *horizontal*
Z: *Z axis/axial flow, mostly referring to the alternative abbreviation for the Saxo turbine type*
S: *S-type (usually referring to Kaplan S-type turbine)*
Bulb: *bulb type (a generator or gear is placed in the bulb)*
Pit: *pit type (generator is placed in the pit)*
ST/TM: *suction turbine/turbine modular*

Runner dimensions

The number coming after the turbine usually refers to the ***runner diameter***

(e.g., KV300: Kaplan-Vertical-300 mm runner diameter)

Gearbox arrangements

D: *direct drive (no gearbox)*
P: *parallel gearbox*
B: *belt transmission*

Conclusion

For anyone working on the design, planning, construction, operation, and maintenance of hydropower facilities, it is vital that they understand and use the same terms and definitions. Using the same language as everyone else enables efficient communication with community members (engineers, operators, regulatory agencies, researchers, etc.). This minimizes the risk of miscommunication, which can lead to technical issues, safety concerns, or regulatory violations.

A discipline such as hydropower is highly multidisciplinary, with civil, mechanical, electrical, environmental, and hydraulic engineering coming together across various projects and at every phase of development. At every level of development, it is necessary to use a common language among those involved in all phases of a project.

This glossary aims to provide a foundational, practical reference to support the technical accuracy and operational efficiency of the hydropower industry and to establish a common technical terminology consistent with current, recognized hydropower industry standards, guidelines, and best practices. While this glossary does not aim to be comprehensive, it provides a solid basis for defining and understanding key terms and technical expressions encountered in design documents, operating manuals, environmental assessments, and contracts.

As hydropower technology continues to change rapidly through advances in turbine designs, developments in digital monitoring systems, the development of more environmentally effective means of mitigating environmental impacts from hydropower plants and increasing integration of hydropower in the flexible and hybrid energy systems, the terminology will continue to grow and change as well. It is critical that we stay abreast of the definitions of these evolving terms so we can encourage innovation, develop sustainably, and support hydropower's continued role as a reliable and responsible contributor to the global transition to low-carbon energy systems.

List of abbreviations

AC	Alternating Current
AST	Archimedes Screw Turbine
EPRI	Electric Power Research Institute
GW	Gigawatt (used to describe global power storage capacity)
PSHPP	Pumped Storage Hydropower Plants
RES	Renewable Energy Source
SETUR	Precession Rolling Turbine (also referred to as bladeless PROTUR)
VSC PSH	Variable Speed Control Pumped Storage Hydropower plant

References

Device for lifting water. (2012). Government Printing Office, *The Official Gazette of the United States Patent Office (Washington, DC: Government Printing Office, 1895)*. Retrieved December 16, 2025, https://etc.usf.edu/clipart/76700/76713/76713_water-mill.htm.

Picone, C., Sinagra, M., & Tucciarelli, T. (2025). A new regulation system for Crossflow type turbines for water management industry. *Energy, 328*, 136695. https://doi.org/10.1016/j.energy.2025.136695.

Timilsina, A. B., Mulligan, S., & Bajracharya, T. R. (2018). Water vortex hydropower technology: a state-of-the-art review of developmental trends. *Clean Technologies and Environmental Policy, 20*(8), 1737–1760. https://doi.org/10.1007/s10098-018-1589-0.

10

Epilogue

For many years, hydropower has been a powerful source of positive change for people. Early humans harnessed nature's energy, and today, hydropower plants paly a crucial part in generating power from renewable sources. The book *Hydropower Plants: A Contemporary Guide for Hydropower Plants in the Global Scale* describes the evolution of hydropower worldwide through advances in technology, digitalization, and materials science, in a complex way. Many think of hydropower as old technology; however, because of its role in supporting new renewable energy forms, hydropower will continue to be vital in large-scale energy storage via pumped storage systems. With advances in turbine technology, particularly in its maintenance, as well as digital retrofitting and providing for flexible, resilient low-carbon grids, hydropower developers are responding to global development trends. Countries that are still developing increasingly rely on small to medium-sized hydropower projects to provide a reliable source of baseload electricity. In contrast, more developed nations are working to upgrade and retrofit their existing hydropower systems and upgrade their overall infrastructure. The new law mandates that agencies regulating hydropower must evaluate ecological impacts, assess how facility operation affects other humans and communities located adjacent to facilities, and develop plans to adapt to climate change through sediment management, habitat connectivity, and maintaining acceptable water quality. Also, there is much interest in preserving the historic and cultural significance of hydropower facilities. Hydropower exists between new ideas, responsibility to nature, and preserving the past; therefore, hydropower's continued evolution will greatly influence its future place in the world.

The research on hydropower plants is truly a dynamic field that may have paused at times, but has always been a driver of change throughout history—whether it's human efforts to harness natural forces, early power for colonization, anticipation of the industrial revolution, or the recent surge in decentralized energy sources.

This book, *Hydropower Plants: A Contemporary Guide for Hydropower Plants in the Global Scope*, aims to show that the field of hydropower is anything but static. From an architectural and civil engineering perspective, hydropower plants are among the few

Hydropower Plants. DOI: https://doi.org/10.1016/B978-0-12-824520-0.00008-3

human made structures that are inherently dynamic, as metaphorically described "the nature is literally flowing through them." In fact, recent pressures from new technologies, changing societal expectations, and global energy demands require the hydropower sector to adapt continually. Building on centuries of engineering tradition, numerous examples and variations of technological solutions—despite often being called a mature technology—generally support or cogenerate solutions for emerging technologies capable of harnessing renewable energy sources, while remaining one of the most efficient energy storage options when pumped storage hydropower is considered.

Throughout the past years, hydropower plants have seen large advancement because of a range of factors, including digital evolution in addition to the introduction of AI-based technology solutions (such as predictive maintenance and energy forecasting tools), new materials used in the construction of turbines, advances in engineering processes, improvements to manufacturing processes, and changes to turbine design on an ongoing basis. The development of turbine technology has created opportunities for hydropower projects to help create a flexible energy grid that supports a distributed and resilient future for energy and transportation systems. Also, advances in manufacturing technologies are reviving some turbine types (such as double-action or dual-runner solutions, e.g., the Francis camelback type) that were originally considered inefficient, obsolete, or subject to replacement by a more modern model because they were constrained by the manufacturing technologies of their time.

While new technologies will help create new ways to develop hydropower projects, the use of new technologies is critical for the future of hydropower in order to become successful in the face of a changing energy landscape caused by climate change, decarbonization, and the increasing inclusion of wind and solar as part of a renewable energy portfolio. Recent changes globally in how countries develop hydropower projects have altered both project design and management processes.

Many developing countries are using smaller and medium-sized hydropower systems as baseload power sources, as they expand their renewable energy systems.

Meanwhile, countries in the developed world focus on modernizing their existing facilities by implementing digital solutions to retrofit and improve efficiency, maximizing use, and prolonging operational life. A shared global theme is that hydropower must serve as a source of energy security and remain adaptable as we shift away from fossil fuels toward a carbon-free future.

Laws are constantly being amended. Environmental regulations now require new policies for designing fish-friendly structures, setting ecological release flows, and managing watersheds. Many licensing frameworks emphasize the importance of local community involvement, climate change adaptation, and adaptive management; therefore, engineers and operators must reconsider their traditional business practices. As a result, new regulations support the development and promotion of hydropower that is environmentally, socially, and economically sustainable.

The environmental concerns linked to the hydropower industry today stem from technological advancements in the field and pose a challenge that will persist for years: integrating hydropower generation with the preservation of river ecosystems. As hydropower continues to develop technologically and new policies are implemented, there will be a greater need for environmental protection, including sediment management, river habitat connectivity, aquatic biodiversity, and water quality control. As climate change impacts water resources more severely, the combined efforts of all stakeholders will be essential to address the environmental challenges of hydropower as a steady source of clean energy.

The recognition of hydropower's industrial history is one of several emerging themes of the 21st century. Many historic hydropower stations, which are also engineering landmarks, are now being acknowledged not just as old infrastructure but also as a specific cultural and industrial heritage, and technological resources, often directly connected to local or regional development.

By preserving and reusing these sites, we merge technological and historic approaches to keep communities connected to their energy heritage and expansion. These developments reflect larger trends in industrial archaeology, the strengthening of regional identity, and the sustainable redevelopment of our communities, and provide an opportunity for hydropower to be viewed as more than simply a power supply.

All these changes clearly indicate that industry is in transition, striving to reconcile past legacies with a future-oriented commitment to embracing new technology. Hydropower now sits at the crossroads of engineering innovation, environmental responsibility, and cultural resource preservation; moving forward, it has a unique chance to bring these three responsibilities into balance, acting as a connecting factor. Therefore, this book recognizes that hydropower's story is ongoing. As we continue our research, engineering design, and policy-making efforts in collaboration with other hydropower stakeholders, hydropower will remain a vital part of the global energy system as an adaptable and evolving resource for the future.

Index

Note: Page numbers followed by "*f*" and "*t*" refer to figures and tables, respectively.

Printed and bound by CPI Group (UK) Ltd, Croydon, CR0 4YY
11/09/2026
02215550-0001